AF556207

FIRST COURSE IN PROBABILITY

FIRST COURSE IN PROBABILITY

By
A.K. Sharma

DISCOVERY PUBLISHING HOUSE PVT. LTD.
NEW DELHI-110 002

First Published - 2009

Reprinted - 2018

ISBN: 978-81-8356-450-2

First Course in Probability

Published by:

DISCOVERY PUBLISHING HOUSE PVT. LTD.
4383/4B, Ansari Road, Darya Ganj
New Delhi-110 002 (India)
Phone: +91-11-23279245, 43596064-65
Fax: +91-11-23253475
E-mail: discoverypublishinghouse@gmail.com
sales@discoverypublishinggroup.com
web: www.discoverypublishinggroup.com

Printed at:
Infinity Imaging Systems
Delhi

Content

Preface

The exact meaning of the word Probability is possibility or chance, that may happen. The theory of probability is used in areas like mathematics, statistics, science and philosophy to make conclusions about the chance of an event. It does not have any definite definition, but there are two broad categories of probability interpreters. First, *Frequentists,* who consider probability to be the relative frequency in the long run of outcomes. Second, *Bayesians,* according to whom probability is the way to represent an individual's degree of belief in a statement, given the evidence.

An important application of this theory in day-to-day life is 'reliability'. Many consumer products, utilise reliability theory in the design of the product in order to reduce the probability of failure. The failure of probability may be closely associated, with the product's warranty. It is widely recognised that the term *probability* is something, which is used in the contexts, where it has nothing to do with physical randomness. The theory of *probability* is one interpretation of the concept of probability. This kind of objective probability is sometimes also called 'Chance'. The main challenge, facing probability theories, is to say exactly, what is meant. At present, unfortunately, none of the well-recognised accounts of propensity comes close to meet this challenge.

This book is to encounter a very genuine need to obtain the information required, in order to reach the assumed level of understanding. Expectedly, this effort would benefit the concerned readers, in serving their purpose. Enlightening comments and feedback is solicited.

Distributions and Densities

Important Distributions

In this chapter, we describe the discrete probability distributions and the continuous probability densities that occur most often in the analysis of experiments. We will also show how one simulates these distributions and densities on a computer.

Discrete Uniform Distribution

In many cases, we assume that all outcomes of an experiment are equally likely. If X is a random variable which represents the outcome of an experiment of this type, then we say that X is uniformly distributed. If the sample space S is of size n, where $0 < n < \infty$, then the distribution function $m(w)$ is defined to be $1/n$ for all $w \in S$. As is the case with all of the discrete probability distributions discussed in this chapter, this experiment can be simulated on a computer using the programme *GeneralSimulation*. However, in this case, a faster algorithm can be used instead. The expression

$$1 + \lfloor n(rnd) \rfloor$$

takes on as a value each integer between 1 and n with probability $1/n$ (the notation $\lfloor x \rfloor$ denotes the greatest integer not exceeding x). Thus, if the possible outcomes of the experiment are labelled

$w_1\ w_2, \ldots, w_n$, then we use the above expression to represent the subscript of the output of the experiment.

If the sample space is a countably infinite set, such as the set of positive integers, then it is not possible to have an experiment which is uniform on this set. If the sample space is an uncountable set, with positive, finite length, such as the interval [0, 1], then we use continuous density functions.

Binomial Distribution

It is the distribution of the random variable which counts the number of heads which occur when a coin is tossed n times, assuming that on any one toss, the probability that a head occurs is p. The distribution function is given by the formula

$$b(n,p,k) = \binom{n}{k} p^k q^{n-k},$$

where $q = 1 - p$.

One straightforward way to simulate a binomial random variable X is to compute the sum of n independent 0 – 1 random variables, each of which take on the value 1 with probability p. This method requires n calls to a random number generator to obtain one value of the random variable. When n is relatively large (say at least 30), the Central Limit Theorem implies that the binomial distribution is well-approximated by the corresponding normal density function with parameters $\mu = np$ and $\sigma = \sqrt{npq}$. Thus, in this case we can compute a value Y of a normal random variable with these parameters, and if $-1/2 \le Y < n + 1/2$, we can use the value

$$\lfloor Y + 1/2 \rfloor$$

to represent the random variable X. If $Y < -1/2$ or $Y > n + 1/2$, we reject Y and compute another value.

Geometric Distribution

Consider a Bernoulli trials process continued for an infinite number of trials; for example, a coin tossed an infinite sequence of times. Thus, we can determine the distribution for any random variable X relating to the experiment provided $P\ (X = a)$ can be computed in terms of a finite number of trials. For example, let T be the number of trials up to and including the first success. Then

$$P(T = 1) = p,$$
$$P(T = 2) = qp,$$
$$P(T = 3) = q^2p,$$

and in general,

$$P(T = n) = q^{n-1}p.$$

To show that this is a distribution, we must show that

$$p + qp + q^2p + \dots = 1.$$

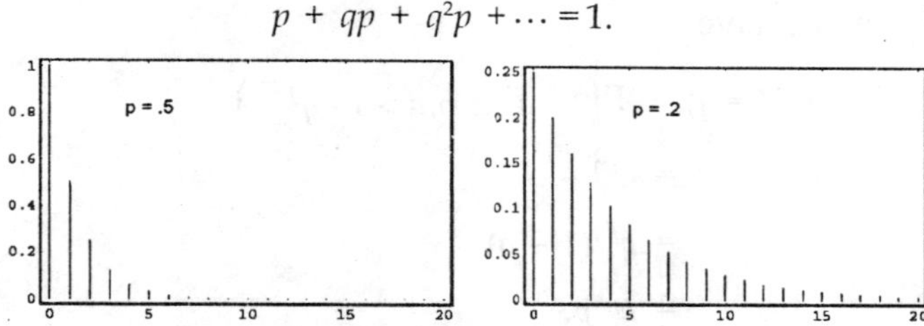

Figure: Geometric distributions.

The left-hand expression is just a geometric series with first term p and common ratio q, so its sum is

$$\frac{p}{1-q}$$

which equals 1.

In the figure above we have plotted this distribution using the programme *GeometricPlot* for the cases $p = .5$ and $p = .2$. We see that as p decreases we are more likely to get large values for T, as would be expected. In both cases, the most probable value for T is 1. This will always be true since

$$\frac{P(T = j+1)}{P(T = j)} = q < 1.$$

In general, if $0 < p < 1$, and $q = 1 - p$, then we say that the random variable T has a geometric distribution if

$$P(T = j) = q^{j-1}p,$$

for $j = 1, 2, 3, \ldots$.

To simulate the geometric distribution with parameter p, we can simply compute a sequence of random numbers in [0, 1), stopping when an entry does not exceed p. However, for small values of p, this is time-consuming (taking, on the average, $1/p$ steps). We now describe a method whose running time does not depend upon the size of p. Define Y to be the smallest integer satisfying the inequality

$$1 - q^Y \geq rnd. \tag{1.1}$$

Then we have

$$\begin{aligned} P(Y = j) &= P\left(1 - q^j \geq rnd > 1 - q^{j-1}\right) \\ &= q^{j-1} - q^j \\ &= q^{j-1}(1 - q) \\ &= q^{j-1}p. \end{aligned}$$

Thus, Y is geometrically distributed with parameter p. To generate Y, all we have to do is solve equation 1.1 for Y. We obtain

$$Y = \left\lceil \frac{\log(1 - rnd)}{\log q} \right\rceil,$$

where the notation $\lceil x \rceil$ means the least integer which is greater than or equal to x. Since $\log(1 - rnd)$ and $\log(rnd)$ are identically distributed, Y can also be generated using the equation

$$Y = \left\lceil \frac{\log rnd}{\log q} \right\rceil.$$

Example: The geometric distribution plays an important role in the theory of queues, or waiting lines. For example, suppose a line of customers waits for service at a counter. It is often assumed that, in each small time unit, either 0 or 1 new customers arrive at the counter. The probability that a customer arrives is p and that no customer arrives is $q = 1 - p$. Then the time T until the next arrival has a geometric distribution. It is natural to ask for the probability that no customer arrives in the next k time units, that is, for $P(T > k)$. This is given by

$$P(T > k) = \sum_{j=k+1}^{\infty} q^{j-1}p = q^k\left(p + qp + q^2p + \cdots\right)$$
$$= q^k .$$

This probability can also be found by noting that we are asking for no successes (i.e., arrivals) in a sequence of k consecutive time units, where the probability of a success in any one time unit is p. Thus, the probability is just q^k, since arrivals in any two time units are independent events.

It is often assumed that the length of time required to service a customer also has a geometric distribution but with a different value for p. This implies a rather special property of the service time. To see this, let us compute the conditional probability

$$P(T > r + s \mid T > r) = \frac{P(T > r+s)}{P(T > r)} = \frac{q^{r+s}}{q^r} = q^s .$$

Thus, the probability that the customer's service takes s more time units is independent of the length of time r that the customer has already been served. Because of this interpretation, this property is called the "memoryless" property, and is also obeyed by the exponential distribution. (Fortunately, not too many service stations have this property.)

Negative Binomial Distribution

Suppose we are given a coin which has probability p of coming up heads when it is tossed. We fix a positive integer k, and toss the coin until the kth head appears. We let X represent the number of tosses. When $k = 1$, X is geometrically distributed. For a general k, we say that X has a negative binomial distribution. We now calculate the probability distribution of X. If $X = x$, then it must be true that there were exactly $k - 1$ heads thrown in the first

$x - 1$ tosses, and a head must have been thrown on the xth toss. There are

$$\binom{x-1}{k-1}$$

sequences of length x with these properties, and each of them is assigned the same probability, namely

$$p^{k-1} q^{x-k}.$$

Therefore, if we define

$$u(x, k, p) = P(X = x),$$

then

$$u(x, k, p) = \binom{x-1}{k-1} p^k q^{x-k}.$$

One can simulate this on a computer by simulating the tossing of a coin. The following algorithm is, in general, much faster. We note that X can be understood as the sum of k outcomes of a geometrically distributed experiment with parameter p. Thus, we can use the following sum as a means of generating X:

$$\sum_{j=1}^{k} \left\lceil \frac{\log rnd_j}{\log q} \right\rceil.$$

Example: A fair coin is tossed until the second time a head turns up. The distribution for the number of tosses is $u(x, 2, p)$. Thus the probability that x tosses are needed to obtain two heads is found by letting $k = 2$ in the above formula. We obtain

$$u(x, 2, 1/2) = \binom{x-1}{1} \frac{1}{2^x},$$

for $x = 2, 3, \ldots$.

In the following figure we give a graph of the distribution for $k = 2$ and $p = .25$. Note that the distribution is quite asymmetric, with a long tail reflecting the fact that large values of x are possible.

Poisson Distribution

The Poisson distribution arises in many situations. It is safe to say that it is one of the three most important discrete probability distributions (the other two being the uniform and the binomial distributions). The Poisson distribution can be viewed as arising from the binomial distribution or from the exponential density.

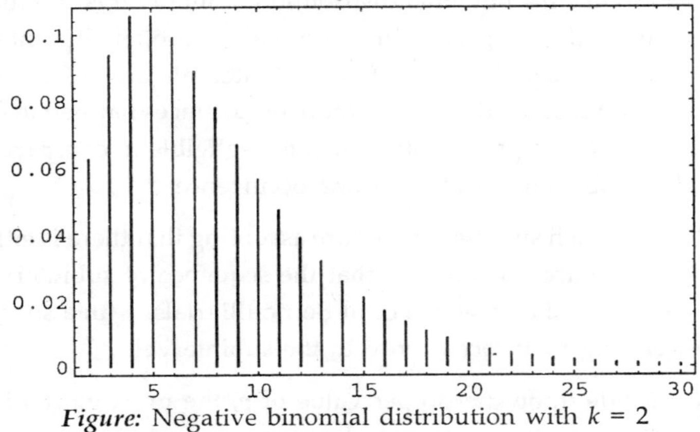

Figure: Negative binomial distribution with $k = 2$ and $p = .25$.

Suppose that we have a situation in which a certain kind of occurrence happens at random over a period of time. For example, the occurrences that we are interested in might be incoming telephone calls to a police station in a large city. We want to model this situation so that we can consider the probabilities of events such as more than 10 phone calls occurring in a 5-minute time interval. Presumably, in our example, there would be more incoming calls between 6:00 and 7:00 pm than between 4:00 and 5:00 am, and this fact would certainly affect the above probability. Thus, to have a hope of computing such probabilities, we must assume that the average rate, i.e., the average number of occurrences per minute, is a constant. This rate we will denote by λ. (Thus, in a given 5-minute time interval, we would expect about 5λ occurrences.) This means that if we were to apply our model to the two time periods given above, we would simply use different rates for the two time periods, thereby obtaining two different probabilities for the given event.

Our next assumption is that the number of occurrences in two non-overlapping time intervals are independent. In our example, this means that the events that there are j calls between 5:00 and 5:15 pm and k calls between 6:00 and 6:15 pm on the same day are independent.

We can use the binomial distribution to model this situation. We imagine that a given time interval is broken up into n subintervals of equal length. If the subintervals are sufficiently short, we can assume that two or more occurrences happen in one subinterval with a probability which is negligible in comparison with the probability of at most one occurrence.

Thus, in each subinterval, we are assuming that there is either 0 or 1 occurrence. This means that the sequence of subintervals can be thought of as a sequence of Bernoulli trials, with a success corresponding to an occurrence in the subinterval.

To decide upon the proper value of p, the probability of an occurrence in a given subinterval, we reason as follows. On the average, there are λ t occurrences in a time interval of length t. If this time interval is divided into n subintervals, then we would expect, using the Bernoulli trials interpretation, that there should be np occurrences. Thus, we want

$$\lambda \mathrm{t} = np,$$

so

$$p = \frac{\lambda t}{n}$$

We now wish to consider the random variable X, which counts the number of occurrences in a given time interval. We want to calculate the distribution of X. For ease of calculation, we will assume that the time interval is of length 1. We know that

$$P(X = 0) = b(n,p,0) = (1-p)^n = \left(1-\frac{\lambda}{n}\right)^n.$$

For large n, this is approximately $e^{-\lambda}$. It is easy to calculate that for any fixed k, we have

$$\frac{b(n,p,k)}{b(n,p,k-1)} = \frac{\lambda-(k-1)p}{kq}$$

which, for large n (and therefore small p) is approximately λ/k. Thus, we have

$$P(X = 1) \approx \lambda e^{-\lambda},$$

and in general,

$$P(X = k) \approx \frac{\lambda^{k}}{k!}e^{-\lambda}. \tag{1.2}$$

The above distribution is the Poisson distribution. We note that it must be checked that the distribution given in equation 1.2 really is a distribution, i.e., that its values are non-negative and sum to 1.

The Poisson distribution is used as an approximation to the binomial distribution when the parameters n and p are large and small, respectively. However, the Poisson distribution also arises in situations where it may not be easy to interpret or measure the parameters n and p.

Example: A typesetter makes, on the average, one mistake per 1000 words. Assume that he is setting a book with 100 words to a page. Let S_{100} be the number of mistakes that he makes on a single page. Then the exact probability distribution for S_{100} would be obtained by considering S_{100} as a result of 100 Bernoulli trials with $p = 1/1000$. The expected value of S_{100} is $\lambda = 100(1/1000) = .1$. The exact probability that $S_{100} = j$ is $b(100, 1/1000, j)$, and the Poisson approximation is

$$\frac{e^{-.1}(.1)^{j}}{j!}.$$

In the following table we give, for various values of n and p, the exact values computed by the binomial distribution and the Poisson approximation.

Table: Poisson approximation to the binomial distribution.

j	*Poisson* $\lambda = .1$	*Binomial* $n = 100$ $p = .001$	*Poisson* $\lambda = 1$	*Binomial* $n = 100$ $p = .01$	*Poisson* $\lambda = 10$	*Binomial* $n = 1000$ $p = .01$
0	.9048	.9048	.3679	.3660	.0000	.0000
1	.0905	.0905	.3679	.3697	.0005	.0004
2	.0045	.0045	.1839	.1849	.0023	.0022
3	.0002	.0002	.0613	.0610	.0076	.0074
4	.0000	.0000	.0153	.0149	.0189	.0186
5			.0031	.0029	.0378	.0374
6			.0005	.0005	.0631	.0627
7			.0001	.0001	.0901	.0900
8			.0000	.0000	.1126	.1128
9					.1251	.1256
10					.1251	.1257
11					.1137	.1143
12					.0948	.0952
13					.0729	.0731
14					.0521	.0520
15					.0347	.0345
16					.0217	.0215
17					.0128	.0126
18					.0071	.0069
19					.0037	.0036
20					.0019	.0018
21					.0009	.0009
22					.0004	.0004
23					.0002	.0002
24					.0001	.0001
25					.0000	.0000

Example: In his book, Feller discusses the statistics of flying bomb hits in the south of London during the Second World War.

Assume that you live in a district of size 10 blocks by 10 blocks so that the total district is divided into 100 small squares. How likely is it that the square in which you live will receive no hits if the total area is hit by 400 bombs?

We assume that a particular bomb will hit your square with probability 1/100. Since there are 400 bombs, we can regard the number of hits that your square receives as the number of successes in a Bernoulli trials process with $n = 400$ and $p = 1{=}100$. Thus we can use the Poisson distribution with $\lambda = 400\ .1/100 = 4$ to approximate the probability that your square will receive j hits. This probability is $p(j) = e^{-4}\ 4^j/j!$. The expected number of squares that receive exactly j hits is then $100 \cdot p(j)$. It is easy to write a programme *LondonBombs* to simulate this situation and compare the expected number of squares with j hits with the observed number.

In the following figure, we have shown the simulated hits, together with a spike graph showing both the observed and predicted frequencies. The observed frequencies are shown as squares, and the predicted frequencies are shown as dots.

If the reader would rather not consider flying bombs, he is invited to instead consider an analogous situation involving cookies and raisins. We assume that we have made enough cookie dough for 500 cookies. We put 600 raisins in the dough, and mix it thoroughly. One way to look at this situation is that we have 500 cookies, and after placing the cookies in a grid on the table, we throw 600 raisins at the cookies.

Example: Suppose that in a certain fixed amount A of blood, the average human has 40 white blood cells. Let X be the random variable which gives the number of white blood cells in a random sample of size A from a random individual. We can think of X as binomially distributed with each white blood cell in the body representing a trial. If a given white blood cell turns up in the sample, then the trial corresponding to that blood cell was a success. Then p should be taken as the ratio of A to the total amount of blood in the individual, and n will be the number of white blood cells in the individual. Of course, in practice, neither of these parameters is very easy to measure accurately, but presumably the number 40 is easy to measure. But for the average human, we then have 40 = np, so we can think of X as being Poisson distributed, with parameter $\lambda = 40$. In this case, it is easier to model the situation using the Poisson distribution than the binomial distribution.

To simulate a Poisson random variable on a computer, a good way is to take advantage of the relationship between the Poisson distribution and the exponential density.

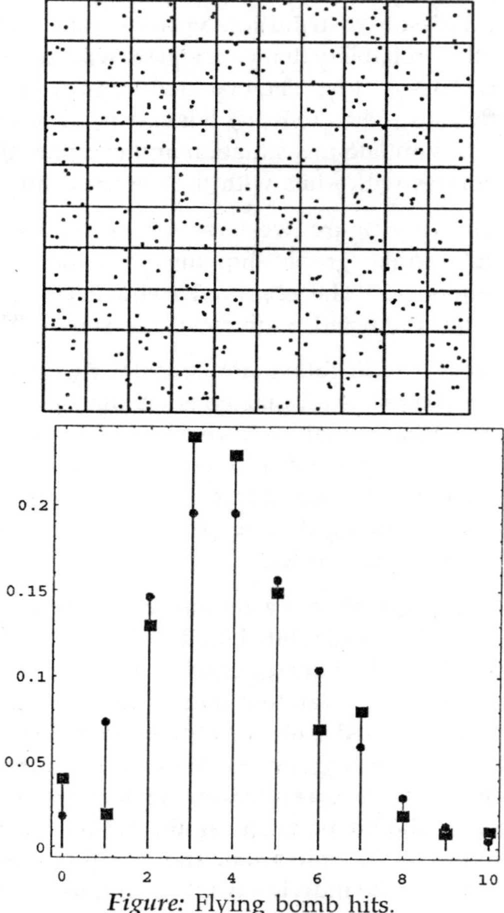

Figure: Flying bomb hits.

Hypergeometric Distribution

Suppose that we have a set of N balls, of which k are red and $N - k$ are blue. We choose n of these balls, without replacement, and define X to be the number of red balls in our sample. The distribution of X is called the hypergeometric distribution. We note that this distribution depends upon three parameters, namely N, k, and n. There does not seem to be a standard notation for this distribution; we will use the notation $h(N, k, n, x)$ to denote $P(X = x)$. This probability can be found by noting that there are

$$\binom{N}{n}$$

different samples of size n, and the number of such samples with exactly x red balls is obtained by multiplying the number of ways of choosing x red balls from the set of k red balls and the number of ways of choosing $n - x$ blue balls from the set of $N - k$ blue balls. Hence, we have

$$h(N,k,n,x) = \frac{\binom{k}{x}\binom{N-k}{n-x}}{\binom{N}{n}}.$$

This distribution can be generalised to the case where there are more than two types of objects.

If we let N and k tend to ∞, in such a way that the ratio k/=N remains fixed, then the hypergeometric distribution tends to the binomial distribution with parameters n and $p = k/N$. This is reasonable because if N and k are much larger than n, then whether we choose our sample with or without replacement should not affect the probabilities very much, and the experiment consisting of choosing with replacement yields a binomially distributed random variable.

Example: It is often of interest to consider two traits, such as eye colour and hair colour, and to ask whether there is an association between the two traits. Two traits are associated if knowing the value of one of the traits for a given person allows us to predict the value of the other trait for that person. The stronger the association, the more accurate the predictions become. If there is no association between the traits, then we say that the traits are independent. In this example, we will use the traits of gender and political party, and we will assume that there are only two possible genders, female and male, and only two possible political parties, Democratic and Republican.

Suppose that we have collected data concerning these traits. To test whether there is an association between the traits, we first assume that there is no association between the two traits. This gives rise to an "expected" data set, in which knowledge of the value of one trait is of no help in predicting the value of the other trait. Our collected data set usually differs from this expected data set. If it differs by quite a bit, then we would tend to reject the assumption of independence of the traits.

To nail down what is meant by "quite a bit," we decide which possible data sets differ from the expected data set by at least as much as ours does, and then we compute the probability that any of these data sets would occur under the assumption of independence of traits. If this probability is small, then it is unlikely that the difference between our collected data set and the expected data set is due entirely to chance.

Table: Observed Data.

	Democrat	*Republican*	
Female	24	4	28
Male	8	14	22
Total	32	18	50

Table: General Data Table.

	Democrat	*Republican*	
Female	s_{11}	s_{12}	t_{11}
Male	s_{21}	s_{22}	t_{12}
	t_{21}	t_{22}	n

Suppose that we have collected the data shown in the previous table. The row and column sums are called marginal totals, or marginals. In what follows, we will denote the row sums by t_{11} and t_{12}, and the column sums by t_{21} and t_{22}.

The ijth entry in the table will be denoted by s_{ij}. Finally, the size of the data set will be denoted by n. Thus, a general data table will look as shown in the previous table. We now explain the model which will be used to construct the "expected" data set. In the model, we assume that the two traits are independent.

We then put t_{21} yellow balls and t_{22} green balls, corresponding to the Democratic and Republican marginals, into an urn. We draw t_{11} balls, without replacement, from the urn, and call these balls females. The t_{12} balls remaining in the urn are called males. In the specific case under consideration, the probability of getting the actual data under this model is given by the expression

$$\frac{\binom{32}{24}\binom{18}{4}}{\binom{50}{28}},$$

i.e., a value of the hypergeometric distribution.

We are now ready to construct the expected data set. If we choose 28 balls out of 50, we should expect to see, on the average, the same percentage of yellow balls in our sample as in the urn. Thus, we should expect to see, on the average, $28(32/50) = 17.92 \approx 18$ yellow balls in our sample.

The other expected values are computed in exactly the same way. Thus, the expected data set is shown in the following table We note that the value of s_{11} determines the other three values in the table, since the marginals are all fixed.

Thus, in considering the possible data sets that could appear in this model, it is enough to consider the various possible values of s_{11}. In the specific case at hand, what is the probability of drawing exactly a yellow balls, i.e., what is the probability that $s_{11} = a$?

Table: Expected Data.

	Democrat	*Republican*	
Female	18	10	28
Male	14	8	22
	32	18	50

It is $\dfrac{\binom{32}{a}\binom{18}{28-a}}{\binom{50}{28}}$. (1.3)

We are now ready to decide whether our actual data differs from the expected data set by an amount which is greater than could be reasonably attributed to chance alone. We note that the expected number of female Democrats is 18, but the actual number in our data is 24. The other data sets which differ from the expected data set by more than ours correspond to those where the number of female Democrats equals 25, 26, 27, or 28. Thus, to obtain the required probability, we sum the expression in (1.3) from $a = 24$ to $a = 28$. We obtain a value of .000395. Thus, we should reject the hypothesis that the two traits are independent.

Finally, we turn to the question of how to simulate a hypergeometric random variable X. Let us assume that the parameters for X are N, k, and n. We imagine that we have a set of N balls, labelled from 1 to N. We decree that the first k of these balls are red, and the rest are blue. Suppose that we have chosen m balls, and that j of them are red. Then there are $k - j$ red balls left, and $N - m$ balls left. Thus, our next choice will be red with probability

$$\frac{k-j}{N-m}.$$

So at this stage, we choose a random number in [0, 1], and report that a red ball has been chosen if and only if the random number does not exceed the above expression. Then we update the values of m and j, and continue until n balls have been chosen.

Benford Distribution

Our next example of a distribution comes from the study of leading digits in data sets. It turns out that many data sets that occur "in real life" have the property that the first digits of the data are not uniformly distributed over the set {1, 2, . . ., 9}. Rather, it appears that the digit 1 is most likely to occur, and that the distribution is monotonically decreasing on the set of possible digits. The Benford distribution appears, in many cases, to fit such data. Many explanations have been given for the occurrence of this distribution. Possibly the most convincing explanation is that this distribution is the only one that is invariant under a change of scale. If one thinks of certain data sets as somehow "naturally occurring," then the distribution should be unaffected by which units are chosen in which to represent the data, i.e., the distribution should be invariant under change of scale.

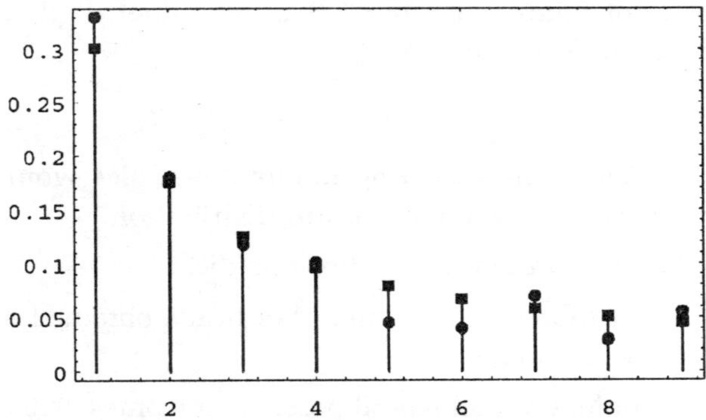

Figure: Leading digits in President Clinton's tax returns.

Theodore Hill gives a general description of the Benford distribution, when one considers the first d digits of integers in a data set. We will restrict our attention to the first digit. In this case, the Benford distribution has distribution function

$$f(k) = \log_{10}(k + 1) - \log_{10}(k),$$

for $1 \leq k \leq 9$.

Mark Nigrini has advocated the use of the Benford distribution as a means of testing suspicious financial records such as bookkeeping entries, checks, and tax returns. His idea is that if someone were to "make up" numbers in these cases, the person would probably produce numbers that are fairly uniformly distributed, while if one were to use the actual numbers, the leading digits would roughly follow the Benford distribution. As an example, Nigrini analysed President Clinton's tax returns for a 13-year period. In the previous figure, the Benford distribution values are shown as squares, and the President's tax return data are shown as circles. One sees that in this example, the Benford distribution fits the data very well.

This distribution was discovered by the astronomer Simon Newcomb who stated the following in his paper on the subject: "That the ten digits do not occur with equal frequency must be evident to anyone making use of logarithm tables, and noticing how much faster the first pages wear out than the last ones. The first significant figure is oftener 1 than any other digit, and the frequency diminishes up to 9.

Exercises

1. For which of the following random variables would it be appropriate to assign a uniform distribution?
   (a) Let X represent the roll of one die.
   (b) Let X represent the number of heads obtained in three tosses of a coin.
   (c) A roulette wheel has 38 possible outcomes: 0, 00, and 1 through 36. Let X represent the outcome when a roulette wheel is spun.
   (d) Let X represent the birthday of a randomly chosen person.
   (e) Let X represent the number of tosses of a coin necessary to achieve a head for the first time.
2. Let n be a positive integer. Let S be the set of integers between 1 and n. Consider the following process: We remove a number from S and write it down. We repeat this until S is empty. The result is a permutation of the integers from 1 to n. Let X denote this permutation. Is X uniformly distributed?

3. Let X be a random variable which can take on countably many values. Show that X cannot be uniformly distributed.
4. Suppose we are attending a college which has 3000 students. We wish to choose a subset of size 100 from the student body. Let X represent the subset, chosen using the following possible strategies. For which strategies would it be appropriate to assign the uniform distribution to X? If it is appropriate, what probability should we assign to each outcome?

   (a) Take the first 100 students who enter the cafeteria to eat lunch.

   (b) Ask the Registrar to sort the students by their Social Security number, and then take the first 100 in the resulting list.

   (c) Ask the Registrar for a set of cards, with each card containing the name of exactly one student, and with each student appearing on exactly one card. Throw the cards out of a third-storey window, then walk outside and pick up the first 100 cards that you find.
5. Under the same conditions as in the preceding exercise, can you describe a procedure which, if used, would produce each possible outcome with the same probability? Can you describe such a procedure that does not rely on a computer or a calculator?
6. Let $X_1, X_2, \ldots, X_n$ be n mutually independent random variables, each of which is uniformly distributed on the integers from 1 to k. Let Y denote the minimum of the X_i's. Find the distribution of Y.
7. A die is rolled until the first time T that a six turns up.

   (a) What is the probability distribution for T?

   (b) Find $P(T > 3)$.

   (c) Find $P(T > 6 \mid T > 3)$.
8. If a coin is tossed a sequence of times, what is the probability that the first head will occur after the fifth toss, given that it has not occurred in the first two tosses?

9. A worker for the Department of Fish and Game is assigned the job of estimating the number of trout in a certain lake of modest size. She proceeds as follows: She catches 100 trout, tags each of them, and puts them back in the lake. One month later, she catches 100 more trout, and notes that 10 of them have tags.
   (a) Without doing any fancy calculations, give a rough estimate of the number of trout in the lake.
   (b) Let N be the number of trout in the lake. Find an expression, in terms of N, for the probability that the worker would catch 10 tagged trout out of the 100 trout that she caught the second time.
   (c) Find the value of N which maximises the expression in part (b). This value is called the maximum likelihood estimate for the unknown quantity N. *Hint*: consider the ratio of the expression for the successive value of N.
10. A census in the United States is an attempt to count everyone in the country. It is inevitable that many people are not counted. The U. S. Census Bureau proposed a way to estimate the number of people who were not counted by the latest census. Their proposal was as follows: In a given locality, let N denote the actual number of people who live there. Assume that the census counted n_1 people living in this area. Now, another census was taken in the locality, and n_2 people were counted. In addition, n_{12} people were counted both times.
    (a) Given N, n_1, and n_2, let X denote the number of people counted both times. Find the probability that $X = k$, where k is a fixed positive integer between 0 and n_2.
    (b) Now assume that $X = n_{12}$. Find the value of N which maximises the expression in part (a). *Hint*: Consider the ratio of the expressions for successive values of N.
11. Suppose that X is a random variable which represents the number of calls coming in to a police station in a one-minute interval. In the text, we showed that X could be modelled using a Poisson distribution with parameter λ, where this

parameter represents the average number of incoming calls per minute. Now suppose that Y is a random variable which represents the number of incoming calls in an interval of length t. Show that the distribution of Y is given by

$$P(Y = k) = e^{-\lambda t} \frac{(\lambda t)^k}{k!},$$

i.e., Y is Poisson with parameter λt. *Hint*: Suppose a Martian were to observe the police station. Let us also assume that the basic time interval used on Mars is exactly t Earth minutes. Finally, we will assume that the Martian understands the derivation of the Poisson distribution in the text. What would she write down for the distribution of Y?

12. Show that the values of the Poisson distribution given in equation 1.2 sum to 1.

13. The Poisson distribution with parameter $\lambda = .3$ has been assigned for the outcome of an experiment. Let X be the outcome function. Find $P(X = 0)$, $P(X = 1)$, and $P(X > 1)$.

14. On the average, only 1 person in 1000 has a particular rare blood type.

    (a) Find the probability that, in a city of 10,000 people, no one has this blood type.

    (b) How many people would have to be tested to give a probability greater than 1/2 of finding at least one person with this blood type?

15. Write a programme for the user to input n, p, j and have the programme print out the exact value of $b(n, p, k)$ and the Poisson approximation to this value.

16. Assume that, during each second, a Dartmouth switchboard receives one call with probability .01 and no calls with probability .99. Use the Poisson approximation to estimate the probability that the operator will miss at most one call if she takes a 5-minute coffee break.

17. The probability of a royal flush in a poker hand is $p = 1/649{,}740$. How large must n be to render the probability of having no royal flush in n hands smaller than $1/e$?

18. A baker blends 600 raisins and 400 chocolate chips into a dough mix and, from this, makes 500 cookies.
    (a) Find the probability that a randomly picked cookie will have no raisins.
    (b) Find the probability that a randomly picked cookie will have exactly two chocolate chips.
    (c) Find the probability that a randomly chosen cookie will have at least two bits (raisins or chips) in it.
19. The probability that, in a bridge deal, one of the four hands has all hearts is approximately 6.3×10^{-12}. In a city with about 50,000 bridge players the resident probability expert is called on the average once a year (usually late at night) and told that the caller has just been dealt a hand of all hearts. Should she suspect that some of these callers are the victims of practical jokes?
20. An advertiser drops 10,000 leaflets on a city which has 2000 blocks. Assume that each leaflet has an equal chance of landing on each block. What is the probability that a particular block will receive no leaflets?
21. In a class of 80 students, the professor calls on 1 student chosen at random for a recitation in each class period. There are 32 class periods in a term.
    (a) Write a formula for the exact probability that a given student is called upon j times during the term.
    (b) Write a formula for the Poisson approximation for this probability. Using your formula estimate the probability that a given student is called upon more than twice.
22. Assume that we are making raisin cookies. We put a box of 600 raisins into our dough mix, mix up the dough, then make from the dough 500 cookies. We then ask for the probability that a randomly chosen cookie will have 0, 1, 2,... raisins. Consider the cookies as trials in an experiment, and let X be the random variable which gives the number of raisins in a given cookie. Then we can regard the number of raisins in a cookie as the result of $n = 600$ independent trials with probability $p = 1/500$ for success on each trial. Since n is large

and p is small, we can use the Poisson approximation with $\lambda = 600(1/500) = 1.2$. Determine the probability that a given cookie will have at least five raisins.

23. For a certain experiment, the Poisson distribution with parameter $\lambda = m$ has been assigned. Show that a most probable outcome for the experiment is the integer value k such that $m - 1 \leq k \leq m$. Under what conditions will there be two most probable values? *Hint:* Consider the ratio of successive probabilities.

24. When John Kemeny was chair of the Mathematics Department at Dartmouth College, he received an average of ten letters each day. On a certain weekday he received no mail and wondered if it was a holiday. To decide this he computed the probability that, in ten years, he would have at least 1 day without any mail. He assumed that the number of letters he received on a given day has a Poisson distribution. What probability did he find? *Hint*: Apply the Poisson distribution twice. First, to find the probability that, in 3000 days, he will have at least 1 day without mail, assuming each year has about 300 days on which mail is delivered.

25. Reese Prosser never puts money in a 10-cent parking metre in Hanover. He assumes that there is a probability of .05 that he will be caught. The first offence costs nothing, the second costs 2 dollars, and subsequent offences cost 5 dollars each. Under his assumptions, how does the expected cost of parking 100 times without paying the metre compare with the cost of paying the metre each time?

Table: Mule Kicks.

Number of deaths	*Number of corps with x deaths in a given year*
0	144
1	91
2	32
3	11
4	2

26. Feller discusses the statistics of flying bomb hits in an area in the south of London during the Second World War. The area in question was divided into $24 \times 24 = 576$ small areas. The total number of hits was 537. There were 229 squares with 0 hits, 211 with 1 hit, 93 with 2 hits, 35 with 3 hits, 7 with 4 hits, and 1 with 5 or more. Assuming the hits were purely random, use the Poisson approximation to find the probability that a particular square would have exactly k hits. Compute the expected number of squares that would have 0, 1, 2, 3, 4, and 5 or more hits and compare this with the observed results.

27. Assume that the probability that there is a significant accident in a nuclear power plant during one year's time is .001. If a country has 100 nuclear plants, estimate the probability that there is at least one such accident during a given year.

28. An airline finds that 4 per cent of the passengers that make reservations on a particular flight will not show up. Consequently, their policy is to sell 100 reserved seats on a plane that has only 98 seats. Find the probability that every person who shows up for the flight will find a seat available.

29. The King's coinmaster boxes his coins 500 to a box and puts 1 counterfeit coin in each box. The King is suspicious, but, instead of testing all the coins in 1 box, he tests 1 coin chosen at random out of each of 500 boxes. What is the probability that he finds at least one fake? What is it if the King tests 2 coins from each of 250 boxes?

30. (From Kemeny) Show that, if you make 100 bets on the number 17 at roulette at Monte Carlo, you will have a probability greater than 1/2 of coming out ahead. What is your expected winning?

31. In one of the first studies of the Poisson distribution, von Bortkiewicz considered the frequency of deaths from kicks in the Prussian army corps. From the study of 14 corps over a 20-year period, he obtained the data shown in the previous table. Fit a Poisson distribution to this data and see if you think that the Poisson distribution is appropriate.

32. It is often assumed that the auto traffic that arrives at the intersection during a unit time period has a Poisson distribution with expected value m. Assume that the number of cars X that arrive at an intersection from the north in unit time has a Poisson distribution with parameter $\lambda = m$ and the number Y that arrive from the west in unit time has a Poisson distribution with parameter $\lambda = \bar{m}$. If X and Y are independent, show that the total number $X + Y$ that arrive at the intersection in unit time has a Poisson distribution with parameter $\lambda = m + \bar{m}$.

33. Cars coming along Magnolia Street come to a fork in the road and have to choose either Willow Street or Main Street to continue. Assume that the number of cars that arrive at the fork in unit time has a Poisson distribution with parameter $\lambda = 4$. A car arriving at the fork chooses Main Street with probability 3/4 and Willow Street with probability 1/4. Let X be the random variable which counts the number of cars that, in a given unit of time, pass by Joe's Barber Shop on Main Street. What is the distribution of X?

34. In the appeal of the People v. Collins case, the counsel for the defence argued as follows: Suppose, for example, there are 5,000,000 couples in the Los Angeles area and the probability that a randomly chosen couple fits the witnesses' description is 1/12,000,000. Then the probability that there are two such couples given that there is at least one is not at all small. Find this probability.

35. A manufactured lot of brass turnbuckles has S items of which D are defective. A sample of s items is drawn without replacement. Let X be a random variable that gives the number of defective items in the sample. Let $p(d) = P(X = d)$.

(a) Show that

$$p(d) = \frac{\binom{D}{d}\binom{S-D}{s-d}}{\binom{S}{s}}$$

Thus, X is hypergeometric.

(b) Prove the following identity, known as Euler's formula:

$$\sum_{d=0}^{\min(D,s)} \binom{D}{d}\binom{S-D}{s-d} = \binom{S}{s}.$$

36. A bin of 1000 turnbuckles has an unknown number D of defectives. A sample of 100 turnbuckles has 2 defectives. The maximum likelihood estimate for D is the number of defectives which gives the highest probability for obtaining the number of defectives observed in the sample. Guess this number D and then write a computer programme to verify your guess.

37. There are an unknown number of moose on Isle Royale (a National Park in Lake Superior). To estimate the number of moose, 50 moose are captured and tagged. Six months later 200 moose are captured and it is found that 8 of these were tagged. Estimate the number of moose on Isle Royale from these data, and then verify your guess by computer programme.

38. A manufactured lot of buggy whips has 20 items, of which 5 are defective. A random sample of 5 items is chosen to be inspected. Find the probability that the sample contains exactly one defective item

    (a) if the sampling is done with replacement.

    (b) if the sampling is done without replacement.

39. Suppose that N and k tend to 1 in such a way that k/N remains fixed. Show that

$$h(N, k, n, x) \to b(n, k/N, x).$$

40. A bridge deck has 52 cards with 13 cards in each of four suits: spades, hearts, diamonds, and clubs. A hand of 13 cards is dealt from a shuffled deck. Find the probability that the hand has

    (a) a distribution of suits 4, 4, 3, 2 (for example, four spades, four hearts, three diamonds, two clubs).

    (b) a distribution of suits 5, 3, 3, 2.

41. Write a computer algorithm that simulates a hypergeometric random variable with parameters N, k, and n.

42. You are presented with four different dice. The first one has two sides marked 0 and four sides marked 4. The second one has a 3 on every side. The third one has a 2 on four sides

and a 6 on two sides, and the fourth one has a 1 on three sides and a 5 on three sides. You allow your friend to pick any of the four dice he wishes. Then you pick one of the remaining three and you each roll your die. The person with the largest number showing wins a dollar. Show that you can choose your die so that you have probability 2/3 of winning no matter which die your friend picks.

43. The students in a certain class were classified by hair colour and eye colour. The conventions used were: Brown and black hair were considered dark, and red and blonde hair were considered light; black and brown eyes were considered dark, and blue and green eyes were considered light. They collected the data shown in the following table. Are these traits independent?

44. Suppose that in the hypergeometric distribution, we let N and *k* tend to 1 in such a way that the ratio k/N approaches a real number *p* between 0 and 1. Show that the hypergeometric distribution tends to the binomial distribution with parameters *n* and *p*.

Table: Observed Data.

	Dark Eyes	**Light Eyes**	
Dark Hair	28	15	43
Light Hair	9	23	32
	37	38	75

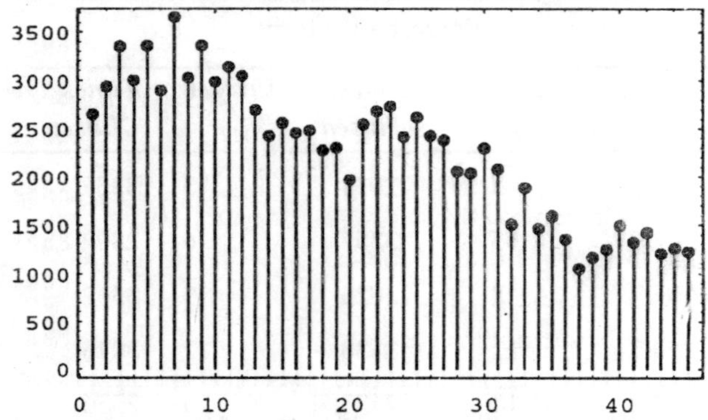

Figure: Distribution of choices in the Powerball lottery.

45 (a) Compute the leading digits of the first 100 powers of 2, and see how well these data fit the Benford distribution.

(b) Multiply each number in the data set of part (a) by 3, and compare the distribution of the leading digits with the Benford distribution.

46. In the Powerball lottery, contestants pick 5 different integers between 1 and 45, and in addition, pick a bonus integer from the same range (the bonus integer can equal one of the first five integers chosen). Some contestants choose the numbers themselves, and others let the computer choose the numbers. The data shown in the table below are the contestant-chosen numbers in a certain state on May 3, 1996. A spike graph of the data is shown in the previous figure.

The goal of this problem is to check the hypothesis that the chosen numbers are uniformly distributed. To do this, compute the value v of the random variable χ^2. In the present case, this random variable has 44 degrees of freedom. The value $v_0 = 59.43$, which represents a number with the property that a χ^2 -distributed random variable takes on values that exceed v_0 only 5 per cent of the time. Does your computed value of v exceed v_0? If so, you should reject the hypothesis that the contestants' choices are uniformly distributed.

Table: Numbers chosen by contestants in the Powerball lottery.

Integer	*Times Chosen*	*Integer*	*Times Chosen*	*Integer*	*Times Chosen*
1	2646	2	2934	3	3352
4	3000	5	3357	6	2892
7	3657	8	3025	9	3362
10	2985	11	3138	12	3043
13	2690	14	2423	15	2556
16	2456	17	2479	18	2276

19	2304	20	1971	21	2543
22	2678	23	2729	24	2414
25	2616	26	2426	27	2381
28	2059	29	2039	30	2298
31	2081	32	1508	33	1887
34	1463	35	1594	36	1354
37	1049	38	1165	39	1248
40	1493	41	1322	42	1423
43	1207	44	1259	45	1224

Important Densities

In this section, we will introduce some important probability density functions give some examples of their use. We will also consider the question of how one simulates a given density using a computer.

Continuous Uniform Density

The simplest density function corresponds to the random variable U whose value represents the outcome of the experiment consisting of choosing a real number at random from the interval $[a, b]$.

$$f(w) = \begin{cases} 1/(b-a), & \text{if } a \le w \le b, \\ 0, & \text{otherwise.} \end{cases}$$

It is easy to simulate this density on a computer. We simply calculate the expression

$$(b - a)rnd + a.$$

Exponential and Gamma Densities

The exponential density function is defined by

$$f(x) = \begin{cases} \lambda e^{-\lambda x}, & \text{if } 0 \le x \le \infty, \\ 0, & \text{otherwise.} \end{cases}$$

Here λ is any positive constant, depending on the experiment. In the following figure we show graphs of several exponential densities for different choices of λ. The exponential density is often used to describe experiments involving a question of the form: How long until something happens?

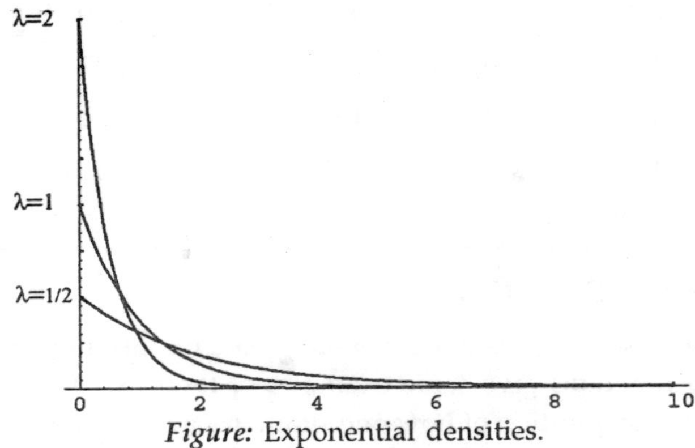

Figure: Exponential densities.

For example, the exponential density is often used to study the time between emissions of particles from a radioactive source.

The cumulative distribution function of the exponential density is easy to compute. Let T be an exponentially distributed random variable with parameter λ. If $x \geq 0$, then we have

$$\begin{aligned} F(x) &= P(T \leq x) \\ &= \int_0^x \lambda e^{-\lambda t}\, dt \\ &= 1 - e^{-\lambda x}. \end{aligned}$$

Both the exponential density and the geometric distribution share a property known as the "memoryless" property. This property says that

$$P(T > r + s \mid T > r) = P(T > s).$$

This can be demonstrated to hold for the exponential density by computing both sides of this equation. The right-hand side is just

$$1 - F(s) = e^{-\lambda s},$$

while the left-hand side is

$$\frac{P(T>r+s)}{P(T>r)} = \frac{1-F(r+s)}{1-F(s)}$$
$$= \frac{e^{-\lambda(r+s)}}{e^{-\lambda r}}$$
$$= e^{-\lambda s}.$$

There is a very important relationship between the exponential density and the Poisson distribution. We begin by defining $X_1, X_2, \ldots$ to be a sequence of independent exponentially distributed random variables with parameter λ. We might think of X_i as denoting the amount of time between the ith and $(i+1)$st emissions of a particle by a radioactive source. (We can think of the parameter λ as representing the reciprocal of the average length of time between emissions. This parameter is a quantity that might be measured in an actual experiment of this type.)

We now consider a time interval of length t, and we let Y denote the random variable which counts the number of emissions that occur in the time interval. We would like to calculate the distribution function of Y (clearly, Y is a discrete random variable). If we let S_n denote the sum $X_1 + X_2 + \ldots + X_n$, then it is easy to see that

$$P(Y = n) = P(S_n \le t \text{ and } S_{n+1} > t).$$

Since the event $S_{n+1} \le t$ is a subset of the event $S_n \le t$, the above probability is seen to be equal to

$$P(S_n \le t) - P(S_{n+1} \le t). \tag{1.4}$$

That the density of S_n is given by the following formula:

$$g_n(x) = \begin{cases} \lambda \dfrac{(\lambda x)^{n-1}}{(n-1)!} e^{-\lambda x}, & \text{if } x > 0 \\ 0, & \text{otherwise} \end{cases}$$

This density is an example of a gamma density with parameters λ and n. The general gamma density allows n to be any positive real number. We shall not discuss this general density.

It is easy to show by induction on n that the cumulative distribution function of S_n is given by:

$$G_n(x) = \begin{cases} 1-e^{-\lambda x}\left(1+\frac{\lambda x}{1!}+\ldots+\frac{(\lambda x)^{n-1}}{(n-1)!}\right), & \text{if } x>0, \\ 0, & \text{otherwise.} \end{cases}$$

Using this expression, the quantity in (1.4) is easy to compute; we obtain

$$e^{-\lambda t}\frac{(\lambda t)^n}{n!},$$

which the reader will recognise as the probability that a Poisson-distributed random variable, with parameter λt, takes on the value n.

The above relationship will allow us to simulate a Poisson distribution, once we have found a way to simulate an exponential density. The following random variable does the job:

$$Y = -\frac{1}{\lambda}\log(rnd). \tag{1.5}$$

Using Corollary 1.2 (below), one can derive the above expression.

We content ourselves for now with a short calculation that should convince the reader that the random variable Y has the required property. We have

$$\begin{aligned} P(Y \le y) &= P\left(-\frac{1}{\lambda}\log(rnd) \le y\right) \\ &= P(\log(rnd) \ge -\lambda y) \\ &= P(rnd \ge e^{-\lambda y}) \\ &= 1-e^{-\lambda y}. \end{aligned}$$

This last expression is seen to be the cumulative distribution function of an exponentially distributed random variable with parameter λ.

To simulate a Poisson random variable W with parameter λ, we simply generate a sequence of values of an exponentially distributed random variable with the same parameter, and keep track of the subtotals S_k of these values. We stop generating the sequence when the subtotal first exceeds λ. Assume that we find that

$$S_n \leq \lambda < S_{n+1}.$$

Then the value n is returned as a simulated value for W.

Example: (Queues) Suppose that customers arrive at random times at a service station with one server, and suppose that each customer is served immediately if no one is ahead of him, but must wait his turn in line otherwise. How long should each customer expect to wait? (We define the waiting time of a customer to be the length of time between the time that he arrives and the time that he begins to be served.)

Let us assume that the interarrival times between successive customers are given by random variables $X_1, X_2, \ldots, X_n$ that are mutually independent and identically distributed with an exponential cumulative distribution function given by

$$F_X(t) = 1 - e^{-\lambda t}.$$

Let us assume, too, that the service times for successive customers are given by random variables $Y_1, Y_2, \ldots, Y_n$ that again are mutually independent and identically distributed with another exponential cumulative distribution function given by

$$F_Y(t) = 1 - e^{-\mu t}.$$

The parameters λ and μ represent, respectively, the reciprocals of the average time between arrivals of customers and the average service time of the customers. Thus, for example, the larger the value of λ, the smaller the average time between arrivals of customers. We can guess that the length of time a customer will spend in the queue depends on the relative sizes of the average interarrival time and the average service time.

It is easy to verify this conjecture by simulation. The programme Queue simulates this queueing process. Let $N(t)$ be the number of customers in the queue at time t.

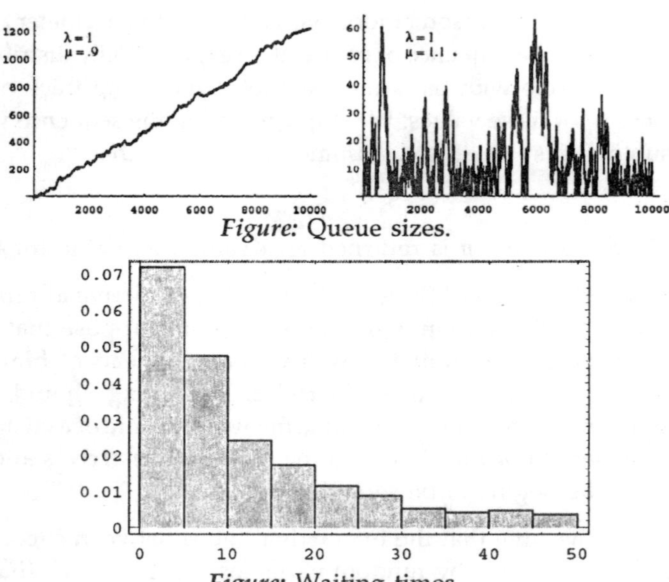

Figure: Queue sizes.

Figure: Waiting times.

Then we plot $N(t)$ as a function of t for different choices of the parameters λ and μ.

We note that when $\lambda < \mu$, then $1/\lambda > 1/\mu$, so the average interarrival time is greater than the average service time, i.e., customers are served more quickly, on average, than new ones arrive. Thus, in this case, it is reasonable to expect that $N(t)$ remains small. However, if $\lambda > \mu$ then customers arrive more quickly than they are served, and, as expected, $N(t)$ appears to grow without limit.

We can now ask: How long will a customer have to wait in the queue for service? To examine this question, we let W_i be the length of time that the ith customer has to remain in the system (waiting in line and being served). Then we can present these data in a bar graph, using the programme Queue, to give some idea of how the W_i are distributed. (Here $\lambda = 1$ and $\mu = 1.1$.)

We see that these waiting times appear to be distributed exponentially. This is always the case when $\lambda < \mu$. The proof of this fact is too complicated to give here, but we can verify it by simulation for different choices of λ and μ, as above.

Functions of a Random Variable

Before continuing our list of important densities, we pause to consider random variables which are functions of other random variables. We will prove a general theorem that will allow us to derive expressions such as equation 1.5.

Theorem: Let X be a continuous random variable, and suppose that $\phi(x)$ is a strictly increasing function on the range of X. Define $Y = \phi(X)$. Suppose that X and Y have cumulative distribution functions F_X and F_Y respectively. Then these functions are related by

$$F_Y(y) = F_X\left(\phi^{-1}(y)\right).$$

If $\phi(x)$ is strictly decreasing on the range of X, then

$$F_Y(y) = 1 - F_X\left(\phi^{-1}(y)\right).$$

Proof: Since ϕ is a strictly increasing function on the range of X, the events $\left(X \le \phi^{-1}(y)\right)$ and $(\phi(X) \le y)$ are equal. Thus, we have

$$\begin{aligned} F_Y(y) &= P(Y \le y) \\ &= P(\phi(X) \le y) \\ &= P\left(X \le \phi^{-1}(y)\right) \\ &= F_X\left(\phi^{-1}(y)\right). \end{aligned}$$

If $\phi(x)$ is strictly decreasing on the range of X, then we have

$$\begin{aligned} F_Y(y) &= P(Y \le y) \\ &= P(\phi(X) \le y) \\ &= P\left(X \ge \phi^{-1}(y)\right) \\ &= 1 - P\left(X \ge \phi^{-1}(y)\right) \\ &= 1 - F_X\left(\phi^{-1}(y)\right). \end{aligned}$$

This completes the proof.

Corollary: Let X be a continuous random variable, and suppose that $\phi(x)$ is a strictly increasing function on the range of X. Define $Y = \phi(X)$. Suppose that the density functions of X and Y are f_X and f_Y, respectively. Then these functions are related by

$$f_Y(y) = f_X\left(\phi^{-1}(y)\right)\frac{d}{dy}\phi^{-1}(y).$$

If $\phi(x)$ is strictly decreasing on the range of X, then

$$f_Y(y) = -f_X\left(\phi^{-1}(y)\right)\frac{d}{dy}\phi^{-1}(y).$$

Proof: This result follows from above theorem by using the Chain Rule.

If the function ϕ is neither strictly increasing nor strictly decreasing, then the situation is somewhat more complicated but can be treated by the same methods. For example, suppose that $Y = X^2$, Then $\phi(x) = x^2$, and

$$\begin{aligned} F_Y(y) &= P(Y \le y) \\ &= P\left(-\sqrt{y} \le X \le +\sqrt{y}\right) \\ &= P\left(X \le +\sqrt{y}\right) - P\left(X \le -\sqrt{y}\right) \\ &= F_X\left(\sqrt{y}\right) - F_X\left(-\sqrt{y}\right). \end{aligned}$$

Moreover,

$$\begin{aligned} f_Y(y) &= \frac{d}{dy}F_Y(y) \\ &= \frac{d}{dy}\left(F_X\left(\sqrt{y}\right) - F_X\left(-\sqrt{y}\right)\right) \\ &= \left(f_X\left(\sqrt{y}\right) + f_X\left(-\sqrt{y}\right)\right)\frac{1}{2\sqrt{y}}. \end{aligned}$$

We see that in order to express F_Y in terms of F_X when $Y = \phi(X)$, we have to express $P(Y \le y)$ in terms of $P(X \le x)$, and this process will depend in general upon the structure of ϕ.

Simulation

Preceding theorem tells us, among other things, how to simulate on the computer a random variable Y with a prescribed cumulative distribution function F. We assume that $F(y)$ is strictly increasing for those values of y where $0 < F(y) < 1$. For this purpose, let U be a random variable which is uniformly distributed on [0, 1]. Then U has cumulative distribution function $F_U(u) = u$. Now, if F is the prescribed cumulative distribution function for Y, then to write Y in terms of U we first solve the equation

$$F(y) = u$$

for y in terms of u. We obtain $y = F^{-1}(u)$. Note that since F is an increasing function this equation always has a unique solution. Then we set $Z = F^{-1}(U)$ and obtain, by previous theorem

$$F_Z(y) = F_U(F(y)) = F(y),$$

since $F_U(u) = u$. Therefore, Z and Y have the same cumulative distribution function. Summarising, we have the following.

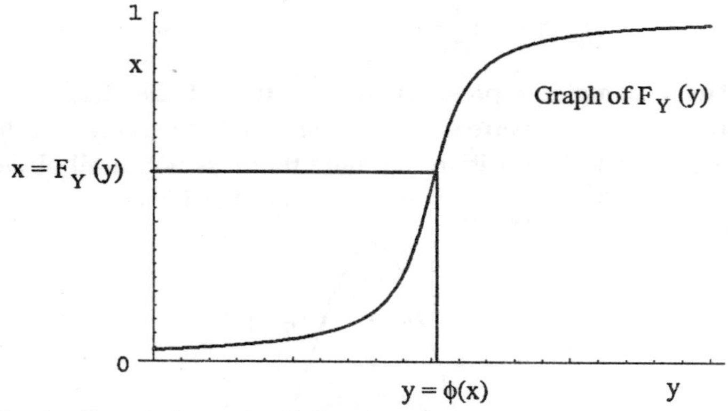

Figure: Converting a uniform distribution F_U into a prescribed distribution F_Y.

Corollary: If $F(y)$ is a given cumulative distribution function that is strictly increasing when $0 < F(y) < 1$ and if U is a random variable with uniform distribution on [0, 1], then

$$Y = F^{-1}(U)$$

has the cumulative distribution $F(y)$.

Thus, to simulate a random variable with a given cumulative distribution F we need only set $Y = F^{-1}(rnd)$.

Normal Density

We now come to the most important density function, the normal density function. The binomial distribution functions are bell-shaped, even for moderate size values of *n*. We recall that a binomially-distributed random variable with parameters *n* and *p* can be considered to be the sum of *n* mutually independent 0-1 random variables. A very important theorem in probability theory, called the Central Limit Theorem, states that under very general conditions, if we sum a large number of mutually independent random variables, then the distribution of the sum can be closely approximated by a certain specific continuous density, called the normal density.

The normal density function with parameters μ and σ is defined as follows:

$$f_X(x) = \frac{1}{\sqrt{2\pi\sigma}} e^{-(x-\mu)^2/2\sigma^2}.$$

The parameter represents the "centre" of the density. The parameter σ is a measure of the "spread" of the density, and thus it is assumed to be positive. We note that it is not at all obvious that the above function is a density, i.e., that its

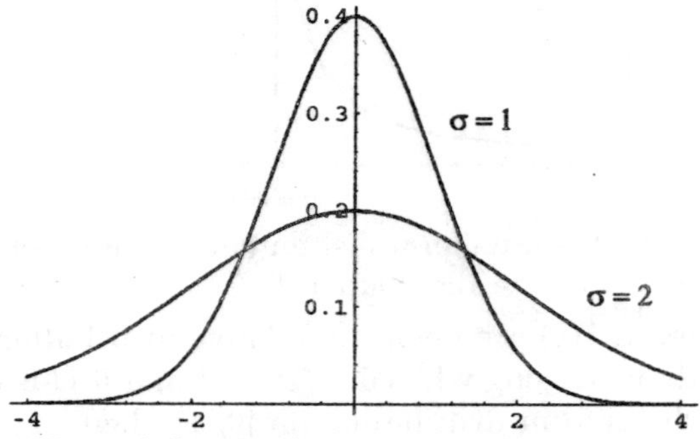

Figure: Normal density for two sets of parameter values.

integral over the real line equals 1. The cumulative distribution function is given by the formula

$$F_X(x) = \int_{-\infty}^{x} \frac{1}{\sqrt{2\pi\sigma}} e^{-(u-\mu)^2/2\sigma^2} du.$$

In the previous figure we have included for comparison a plot of the normal density for the cases $\mu = 0$ and $\sigma = 1$, and $\mu = 0$ and $\sigma = 2$.

One cannot write F_X in terms of simple functions. This leads to several problems. First of all, values of F_X must be computed using numerical integration. Extensive tables exist containing values of this function. Secondly, we cannot write F_X^{-1} in closed form, so we cannot use previous corollary to help us simulate a normal random variable. For this reason, special methods have been developed for simulating a normal distribution. One such method relies on the fact that if U and V are independent random variables with uniform densities on [0, 1], then the random variables

$$X = \sqrt{-2\log U}\cos 2\pi V$$

and

$$Y = \sqrt{-2\log U}\sin 2\pi V$$

are independent, and have normal density functions with parameters $\mu = 0$ and $\sigma = 1$.

Let Z be a normal random variable with parameters $\mu = 0$ and $\sigma = 1$. A normal random variable with these parameters is said to be a standard normal random variable. It is an important and useful fact that if we write

$$X = \sigma Z + \mu,$$

then X is a normal random variable with parameters μ and σ. To show this, we will use previous theorem. We have $\phi(z) = \sigma z + \mu, \phi^{-1}(x) = (x-\mu)/\sigma$, and

$$F_X(x) = F_Z\left(\frac{x-\mu}{\sigma}\right),$$

$$f_X(x) = f_Z\left(\frac{x-\mu}{\sigma}\right)\cdot\frac{1}{\sigma}$$

$$= \frac{1}{\sqrt{2\pi}\sigma} e^{-(x-\mu)^2/2\sigma^2}.$$

The reader will note that this last expression is the density function with parameters μ and σ, as claimed.

We have seen above that it is possible to simulate a standard normal random variable Z. If we wish to simulate a normal random variable X with parameters μ and σ, then we need only transform the simulated values for Z using the equation $X = \sigma Z + \mu$.

Suppose that we wish to calculate the value of a cumulative distribution function for the normal random variable X, with parameters μ and σ. We can reduce this calculation to one concerning the standard normal random variable Z as follows:

$$F_X(x) = P(X \leq x)$$

$$= P\left(Z \leq \frac{x-\mu}{\sigma}\right)$$

$$= F_z\left(\frac{x-\mu}{\sigma}\right).$$

This last expression can be found in a table of values of the cumulative distribution function for a standard normal random variable. Thus, we see that it is unnecessary to make tables of normal distribution functions with arbitrary μ and σ.

The process of changing a normal random variable to a standard normal random variable is known as standardisation. If X has a normal distribution with parameters μ and σ and if

$$Z = \frac{X-\mu}{\sigma},$$

then Z is said to be the standardised version of X.

The following example shows how we use the standardised version of a normal random variable X to compute specific probabilities relating to X.

Example: Suppose that X is a normally distributed random variable with parameters $\mu = 10$ and $\sigma = 3$. Find the probability that X is between 4 and 16.

To solve this problem, we note that $Z = (X - 10)/3$ is the standardised version of X. So, we have

$$\begin{aligned} P(4 \le X \le 16) &= P(X \le 16) - P(X \le 4) \\ &= F_X(16) - F_X(4) \\ &= F_Z\left(\frac{16-10}{3}\right) - F_Z\left(\frac{4-10}{3}\right) \\ &= F_Z(2) - F_Z(-2). \end{aligned}$$

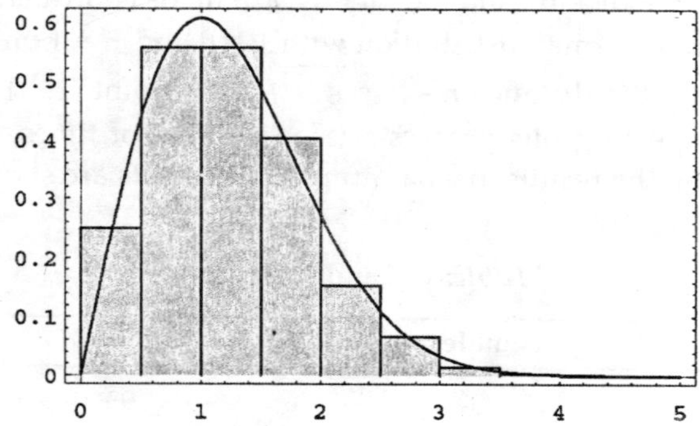

Figure: Distribution of dart distances in 1000 drops.

This last expression can be evaluated by using tabulated values of the standard normal distribution function; when we use this table, we find that $F_Z(2) = .9772$ and $F_Z(-2) = .0228$. Thus, the answer is .9544.

The parameter μ is the mean, or average value, of the random variable X. The parameter σ is a measure of the spread of the random variable, and is called the standard deviation. Thus, the question asked in this example is of a typical type, namely, what is the probability that a random variable has a value within two standard deviations of its average value.

Maxwell and Rayleigh Densities

Example: Suppose that we drop a dart on a large table top, which we consider as the xy-plane, and suppose that the x and

y coordinates of the dart point are independent and have a normal distribution with parameters $\mu = 0$ and $\sigma = 1$. How is the distance of the point from the origin distributed?

This problem arises in physics when it is assumed that a moving particle in R^n has components of the velocity that are mutually independent and normally distributed and it is desired to find the density of the speed of the particle. The density in the case $n = 3$ is called the Maxwell density.

The density in the case $n = 2$ (i.e. the dart board experiment described above) is called the Rayleigh density. We can simulate this case by picking independently a pair of coordinates (x, y), each from a normal distribution with $\mu = 0$ and $\sigma = 1$ on $(-\infty, \infty)$, calculating the distance $r = \sqrt{x^2 + y^2}$ of the point (x, y) from the origin, repeating this process a large number of times, and then presenting the results in a bar graph. The results are shown in the previous figure.

Table: Calculus class data.

	Female	Male	
A	37	56	93
B·	63	60	123
C	47	43	90
Below C	5	8	13
	152	167	319

Table: Expected data.

	Female	Male	
A	44.3	48.7	93
B	58.6	64.4	123
C	42.9	47.1	90
Below C	6.2	6.8	13
	152	167	319

We have also plotted the theoretical density

$$f(r) = re^{-r^2/2}.$$

Chi-Squared Density

It is frequently the case that we have two traits, each of which have several different values. Quite a lot of calculation was needed even in the case of two values for each trait. We now give another method for testing independence of traits, which involves much less calculation.

Example: Suppose that we have the data shown in the previous table concerning grades and gender of students in a Calculus class. We imagine that we have an urn with 319 balls of two colours, say blue and red, corresponding to females and males, respectively. We now draw 93 balls, without replacement, from the urn. These balls correspond to the grade of A. We continue by drawing 123 balls, which correspond to the grade of B. When we finish, we have four sets of balls, with each ball belonging to exactly one set. (We could have stipulated that the balls were of four colours, corresponding to the four possible grades. In this case, we would draw a subset of size 152, which would correspond to the females. The balls remaining in the urn would correspond to the males. The choice does not affect the final determination of whether we should reject the hypothesis of independence of traits).

If we do this, we obtain the expected values shown in the previous table. Even if the traits are independent, we would still expect to see some differences between the numbers in corresponding boxes in the two tables. However, if the differences are large, then we might suspect that the two traits are not independent. We used the probability distribution of the various possible data sets to compute the probability of finding a data set that differs from the expected data set by at least as much as the actual data set does. We could do the same in this case, but the amount of computation is enormous.

Instead, we will describe a single number which does a good job of measuring how far a given data set is from the expected one. To quantify how far apart the two sets of numbers are, we could sum the squares of the differences of the corresponding numbers. (We could also sum the absolute values of the differences, but we would not want to sum the differences.) Suppose that we have data in which we expect to see 10 objects of a certain type, but instead we see 18, while in another case we expect to see 50 objects of a certain type, but instead we see 58. Even though the two differences are about the same, the first difference is more surprising than the second, since the expected number of outcomes in the second case is quite a bit larger than the expected number in the first case. One way to correct for this is to divide the individual squares of the differences by the expected number for that box. Thus, if we label the values in the eight boxes in the first table by O_i (for observed values) and the values in the eight boxes in the second table by E_i (for expected values), then the following expression might be a reasonable one to use to measure how far the observed data is from what is expected:

$$\sum_{i=1}^{8} \frac{(O_i - E_i)^2}{E_i}.$$

This expression is a random variable, which is usually denoted by the symbol χ^2, pronounced "ki-squared." It is called this because, under the assumption of independence of the two traits, the density of this random variable can be computed and is approximately equal to a density called the chi-squared density. We choose not to give the explicit expression for this density, since it involves the gamma function, which we have not discussed. The chi-squared density is, in fact, a special case of the general gamma density.

In applying the chi-squared density, tables of values of this density are used, as in the case of the normal density. The chi-squared density has one parameter n, which is called the number of degrees of freedom. The number n is usually easy to determine from the problem at hand. For example, if we are checking two traits for independence, and the two traits have a and b values, respectively, then the number of degrees of freedom of the random variable χ^2 is $(a - 1)(b - 1)$. So, in the example at hand, the number of degrees of freedom is 3.

We recall that in this example, we are trying to test for independence of the two traits of gender and grades. If we assume these traits are independent, then the ball-and-urn model given above gives us a way to simulate the experiment. Using a computer, we have performed 1000 experiments, and for each one, we have calculated a value of the random variable χ^2. The results are shown in the following figure, together with the chi-squared density function with three degrees of freedom.

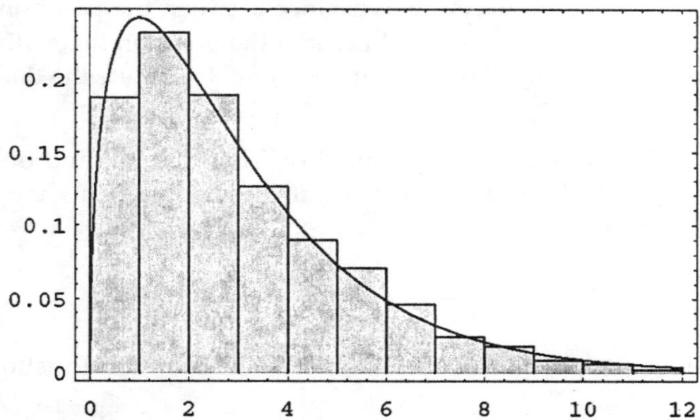

Figure: Chi-squared density with three degrees of freedom.

As we stated above, if the value of the random variable χ^2 is large, then we would tend not to believe that the two traits are independent. But how large is large? The actual value of this random variable for the data above is 4.13. In the previous figure, we have shown the chi-squared density with 3 degrees of freedom. It can be seen that the value 4.13 is larger than most of the values taken on by this random variable.

Typically, a statistician will compute the value v of the random variable χ^2, just as we have done. Then, by looking in a table of values of the chi-squared density, a value v_0 is determined which is only exceeded 5 per cent of the time. If $v \geq v_0$, the statistician rejects the hypothesis that the two traits are independent. In the present case, $v_0 = 7.815$, so we would not reject the hypothesis that the two traits are independent.

Cauchy Density

The following example is from Feller.

Example: Suppose that a mirror is mounted on a vertical axis, and is free to revolve about that axis. The axis of the mirror is 1 foot from a straight wall of infinite length. A pulse of light is shown onto the mirror, and the reflected ray hits the wall. Let ϕ be the angle between the reflected ray and the line that is perpendicular to the wall and that runs through the axis of the mirror. We assume that ϕ is uniformly distributed between $-\pi/2$ and $\pi/2$. Let X represent the distance between the point on the wall that is hit by the reflected ray and the point on the wall that is closest to the axis of the mirror. We now determine the density of X.

Let B be a fixed positive quantity. Then $X \geq B$ if and only if $\tan(\phi) \geq B$, which happens if and only if $\phi \geq \arctan(B)$. This happens with probability

$$\frac{\pi/2 - \arctan(B)}{\pi}$$

Thus, for positive B, the cumulative distribution function of X is

$$F(B) = 1 - \frac{\pi/2 - \arctan(B)}{\pi}$$

Therefore, the density function for positive B is

$$f(B) = \frac{1}{\pi(1+B^2)}.$$

Since the physical situation is symmetric with respect to $\phi = 0$, it is easy to see that the above expression for the density is correct for negative values of B as well.

The Law of Large Numbers states that in many cases, if we take the average of independent values of a random variable, then the average approaches a specific number as the number of values increases. It turns out that if one does this with a Cauchy-distributed random variable, the average does not approach any specific number.

Exercises

1. Choose a number U from the unit interval [0, 1] with uniform distribution. Find the cumulative distribution and density for the random variables
   (a) $Y = U + 2$.
   (b) $Y = U^3$.
2. Choose a number U from the interval [0, 1] with uniform distribution. Findthe cumulative distribution and density for the random variables
   (a) Y = 1=(U + 1).
   (b) Y = log(U + 1).
3. Use previous corollary to derive the expression for the random variable given in equation 1.5. *Hint*: The random variables 1 - *rnd* and *rnd* are identically distributed.
4. Suppose we know a random variable Y as a function of the uniform random variable $U : Y = \phi(U)$, and suppose we have calculated the cumulative distribution function $F_Y(y)$ and thence the density $f_Y(y)$. How can we check whether our answer is correct? An easy simulation provides the answer: Make a bar graph of $Y = \phi(rnd)$ and compare the result with the graph of $f_Y(y)$.These graphs should look similar. Check your answers to Exercises 1 and 2 by this method.
5. Choose a number U from the interval [0, 1] with uniform distribution. Find the cumulative distribution and density for the random variables
   (a) $Y = |U - 1/2|$.
   (b) $Y = (U - 1/2)^2$.
6. Check your results for Exercise 5 by simulation as described in Exercise 4.
7. Explain how you can generate a random variable whose cumulative distribution function is

$$F(x) = \begin{cases} 0, & \text{if } x < 0, \\ x^2, & \text{if } 0 \leq x \leq 1, \\ 1, & \text{if } x > 1. \end{cases}$$

8. Write a programme to generate a sample of 1000 random outcomes each of which is chosen from the distribution given in Exercise 7. Plot a bar graph of your results and compare this empirical density with the density for the cumulative distribution given in Exercise 7.

9. Let U, V be random numbers chosen independently from the interval [0, 1] with uniform distribution. Find the cumulative distribution and density of each of the variables

   (a) $Y = U + V$.

   (b) $Y = |U - V|$.

10. Let U, V be random numbers chosen independently from the interval [0, 1]. Find the cumulative distribution and density for the random variables

    (a) $Y = \max(U, V)$.

    (b) $Y = \min(U, V)$.

11. Write a programme to simulate the random variables of Exercises 9 and 10 and plot a bar graph of the results. Compare the resulting empirical density with the density found in Exercises 9 and 10.

12. A number U is chosen at random in the interval [0, 1]. Find the probability that

    (a) $R = U^2 < 1=4$.

    (b) $S = U(1 - U) < 1/4$.

    (c) $T = U =(1 - U) < 1/4$.

13. Find the cumulative distribution function F and the density function f for each of the random variables R, S, and T in Exercise 12.

14. A point P in the unit square has coordinates X and Y chosen at random in the interval [0, 1]. Let D be the distance from P to the nearest edge of the square, and E the distance to the nearest corner. What is the probability that

    (a) $D < 1/4$?

    (b) $E < 1/4$?

15. In Exercise 14 find the cumulative distribution F and density f for the random variable D.

16. Let X be a random variable with density function

$$f_X(x) = \begin{cases} cx(1-x), & \text{if } 0 < x < 1, \\ 0, & \text{otherwise}. \end{cases}$$

   (a) What is the value of c?

   (b) What is the cumulative distribution function F_X for X?

   (c) What is the probability that $X < 1/4$?

17. Let X be a random variable with cumulative distribution function

$$F(x) = \begin{cases} 0, & \text{if } x > 0, \\ \sin^2(\pi x/2), & \text{if } 0 \le x \le 1, \\ 1, & \text{if } 1 < x. \end{cases}$$

   (a) What is the density function f_X for X?

   (b) What is the probability that $X < 1/4$?

18. Let X be a random variable with cumulative distribution function F_X, and let $Y = X + b$, $Z = aX$, and $W = aX + b$, where a and b are any constants. Find the cumulative distribution functions F_Y, F_Z, and F_W. *Hint*: The cases $a > 0$, $a = 0$, and $a < 0$ require different arguments.

19. Let X be a random variable with density function f_X, and let $Y = X + b$, $Z = aX$, and $W = aX + b$, where $a \neq 0$. Find the density functions f_Y, f_Z, and f_W.

20. Let X be a random variable uniformly distributed over $[c, d]$, and let $Y = aX + b$. For what choice of a and b is Y uniformly distributed over [0, 1]?

21. Let X be a random variable with cumulative distribution function F strictly increasing on the range of X. Let $Y = F(X)$. Show that Y is uniformly distributed in the interval [0, 1]. (The formula $X = F^{-1}(Y)$ then tells us how to construct X from a uniform random variable Y.)

22. Let X be a random variable with cumulative distribution function F. The median of X is the value m for which

$F(m) = 1/2$. Then $X < m$ with probability 1/2 and $X > m$ with probability 1/2. Find m if X is

(a) uniformly distributed over the interval $[a, b]$.

(b) normally distributed with parameters μ and σ.

(c) exponentially distributed with parameter λ.

23. Let X be a random variable with density function f_x. The mean of X is the value $\mu = \int x f_x(x)\,dx$. Then μ gives an average value for X. Find μ if X is distributed uniformly, normally, or exponentially, as in Exercise 22.

Table: Grading on the curve.

Test Score	*Letter grade*
$\mu + \sigma < x$	A
$\mu < x < \mu + \sigma$	B
$\mu - \sigma < x < \mu$	C
$\mu - 2\sigma < x < \mu - \sigma$	D
$x < \mu - 2\sigma$	F

24. Let X be a random variable with density function f_X. The mode of X is the value M for which $f(M)$ is maximum. Then values of X near M are most likely to occur. Find M if X is distributed normally or exponentially, as in Exercise 22. What happens if X is distributed uniformly?

25. Let X be a random variable normally distributed with parameters $\mu = 70, \sigma = 10$. Estimate

(a) $P(X > 50)$.

(b) $P(X < 60)$.

(c) $P(X > 90)$.

(d) $P(60 < X < 80)$.

26. Bridies' Bearing Works manufactures bearing shafts whose diameters are normally distributed with parameters $\mu = 1$, $\sigma = .002$. The buyer's specifications require these diameters to be $1.000 \pm .003$ cm. What fraction of the manufacturer's

shafts are likely to be rejected? If the manufacturer improves her quality control, she can reduce the value of σ. What value of σ will ensure that no more than 1 per cent of her shafts are likely to be rejected?

27. A final examination at Podunk University is constructed so that the test scores are approximately normally distributed, with parameters μ and σ. The instructor assigns letter grades to the test scores as shown in the previous table (this is the process of "grading on the curve").

    What fraction of the class gets A, B, C, D, F?

28. (Ross) An expert witness in a paternity suit testifies that the length (in days) of a pregnancy, from conception to delivery, is approximately normally distributed, with parameters $\mu = 270$, $\sigma = 10$. The defendant in the suit is able to prove that he was out of the country during the period from 290 to 240 days before the birth of the child. What is the probability that the defendant was in the country when the child was conceived?

29. Suppose that the time (in hours) required to repair a car is an exponentially distributed random variable with parameter $\lambda = 1/2$. What is the probability that the repair time exceeds 4 hours? If it exceeds 4 hours what is the probability that it exceeds 8 hours?

30. Suppose that the number of years a car will run is exponentially distributed with parameter $\mu = 1/4$. If Prosser buys a used car today, what is the probability that it will still run after 4 years?

31. Let U be a uniformly distributed random variable on [0, 1]. What is the probability that the equation

$$x^2 + 4Ux + 1 = 0$$

    has two distinct real roots x_1 and x_2?

32. Write a programme to simulate the random variables whose densities are given by the following, making a suitable bar graph of each and comparing the exact density with the bar graph.

(a) $f_X(x) = e^{-x}$ on $[0, \infty)$ (but just do it on [0, 10]).

(b) $f_X(x) = 2x$ on [0, 1].

(c) $f_X(x) = 3x^2$ on [0, 1].

(d) $f_X(x) = 4|x - 1/2|$ on [0, 1].

33. Suppose we are observing a process such that the time between occurrences is exponentially distributed with $\lambda = 1/30$ (i.e., the average time between occurrences is 30 minutes). Suppose that the process starts at a certain time and we start observing the process 3 hours later. Write a programme to simulate this process. Let T denote the length of time that we have to wait, after we start our observation, for an occurrence. Have your programme keep track of T. What is an estimate for the average value of T?

34. Jones puts in two new lightbulbs: a 60 watt bulb and a 100 watt bulb. It is claimed that the lifetime of the 60 watt bulb has an exponential density with average lifetime 200 hours ($\lambda = 1/200$). The 100 watt bulb also has an exponential density but with average lifetime of only 100 hours ($\lambda = 1/100$). Jones wonders what is the probability that the 100 watt bulb will outlast the 60 watt bulb.

If X and Y are two independent random variables with exponential densities $f(x) = \lambda e^{-\lambda x}$ and $g(x) = \mu e^{-\mu x}$, respectively, then the probability that X is less than Y is given by

$$P(X < Y) = \int_0^\infty f(x)(1 - G(x))\,dx,$$

where $G(x)$ is the cumulative distribution function for $g(x)$. Explain why this is the case. Use this to show that

$$P(X < Y) = \frac{\lambda}{\lambda + \mu}$$

and to answer Jones's question.

35. Suppose that you watch the size of the queue. If there are j people in the queue the next time the queue size changes it will either decrease to $j - 1$ or increase to $j + 1$. Use the result of Exercise 34 to show that the probability that the queue size decreases to $j - 1$ is $\mu/(\mu + \lambda)$ and the probability

that it increases to $j+1$ is $\lambda/(\mu+\lambda)$. When the queue size is 0 it can only increase to 1. Write a programme to simulate the queue size. Use this simulation to help formulate a conjecture containing conditions on μ and λ that will ensure that the queue will have times when it is empty.

36. Let X be a random variable having an exponential density with parameter λ. Find the density for the random variable $Y = rX$, where r is a positive real number.

37. Let X be a random variable having a normal density and consider the random variable $Y = e^X$. Then Y has a log normal density. Find this density of Y.

38. Let X_1 and X_2 be independent random variables and for $i = 1, 2$, let $Y_i = \phi_i(X_i)$, where ϕ_i is strictly increasing on the range of X_i. Show that Y_1 and Y_2 are independent. Note that the same result is true without the assumption that the ϕ_i's are strictly increasing, but the proof is more difficult.

Probability Densities

Simulation of Continuous Probabilities

In this chapter we shall show how we can use computer simulations for experiments that have a whole continuum of possible outcomes.

Probabilities

Example: We begin by constructing a spinner, which consists of a circle of unit circumference and a pointer as shown in the following figure. We pick a point on the circle and label it 0, and then label every other point on the circle with the distance, say x, from 0 to that point, measured counterclockwise. The experiment consists of spinning the pointer and recording the label of the point at the tip of the pointer. We let the random variable X denote the value of this outcome. The sample space is clearly the interval $[0, 1)$. We would like to construct a probability model in which each outcome is equally likely to occur.

If we proceed for experiments with a finite number of possible outcomes, then we must assign the probability 0 to each outcome, since otherwise, the sum of the probabilities, overall of the possible outcomes, would not equal 1. (In fact, summing an uncountable number of real numbers is a tricky business; in particular, in order for such a sum to have any meaning, at most countably many of the summands can be different than 0.) However, if all of the assigned probabilities are 0, then the sum is 0, not 1, as it should be.

At present, we will assume that such a model can be constructed. We will also assume that in this model, if E is an arc of the circle, and E is of length p, then the model will assign the probability p to E. This means that if the pointer is spun, the probability that it ends up pointing to a point in E equals p, which is certainly a reasonable thing to expect.

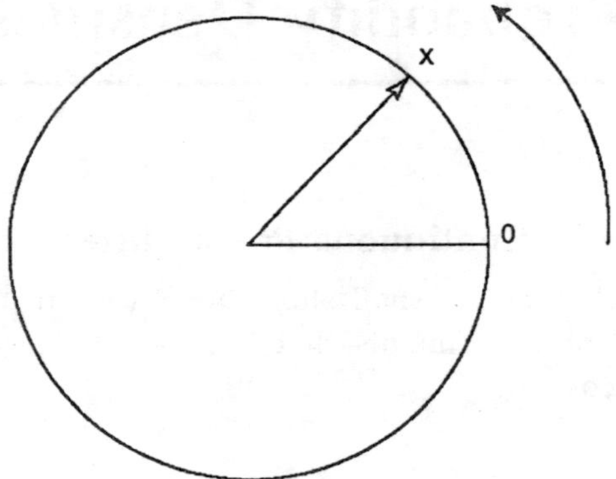

Figure: A spinner.

To simulate this experiment on a computer is an easy matter. Many computer software packages have a function which returns a random real number in the interval [0, 1]. Actually, the returned value is always a rational number, and the values are determined by an algorithm, so a sequence of such values is not truly random. Nevertheless, the sequences produced by such algorithms behave much like theoretically random sequences, so we can use such sequences in the simulation of experiments. On occasion, we will need to refer to such a function. We will call this function *rnd*.

Monte Carlo Procedure and Areas

It is sometimes desirable to estimate quantities whose exact values are difficult or impossible to calculate exactly. In some of these cases, a procedure involving chance, called a Monte Carlo procedure, can be used to provide such an estimate.

Example: In this example we show how simulation can be used to estimate areas of plane figures. Suppose that we programme our computer to provide a pair (x, y) or numbers, each chosen independently at random from the interval [0, 1]. Then we can interpret this pair (x, y) as the coordinates of a point chosen at random from the unit square. Events are subsets of the unit square. Our experience with previous example suggests that the point is equally likely to fall in subsets of equal area. Since the total area of the square is 1, the probability of the point falling in a specific subset E of the unit square should be equal to its area. Thus, we can estimate the area of any subset of the unit square by estimating the probability that a point chosen at random from this square falls in the subset.

We can use this method to estimate the area of the region E under the curve $y = x^2$ in the unit square. We choose a large number of points (x, y) at random and record what fraction of them fall in the region $E = \{(x, y) : y \leq x^2\}$.

The programme *MonteCarlo* will carry out this experiment for us. Running this programme for 10,000 experiments gives an estimate of .325.

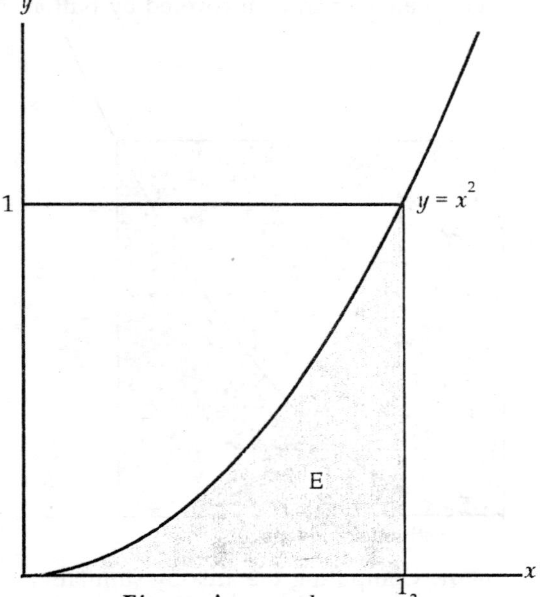

Figure: Area under $y = x^2$

From these experiments we would estimate the area to be about 1/3. Of course, for this simple region we can find the exact area by calculus. In fact,

$$\text{Area of } E = \int_0^1 x^2 dx = \frac{1}{3}.$$

When we simulate an experiment of this type n times to estimate a probability, we can expect the answer to be in error by at most $1/\sqrt{n}$ at least 95 per cent of the time. For 10,000 experiments we can expect an accuracy of 0.01, and our simulation did achieve this accuracy. This same argument works for any region E of the unit square. For example, suppose E is the circle with centre (1/2, 1/2) and radius 1/2. Then the probability that our random point (x, y) lies inside the circle is equal to the area of the circle, that is,

$$P(E) = \pi\left(\frac{1}{2}\right)^2 = \frac{\pi}{4}.$$

If we did not know the value of π, we could estimate the value by performing this experiment a large number of times! The above example is not the only way of estimating the value of π by a chance experiment. Here is another way, discovered by Buffon.

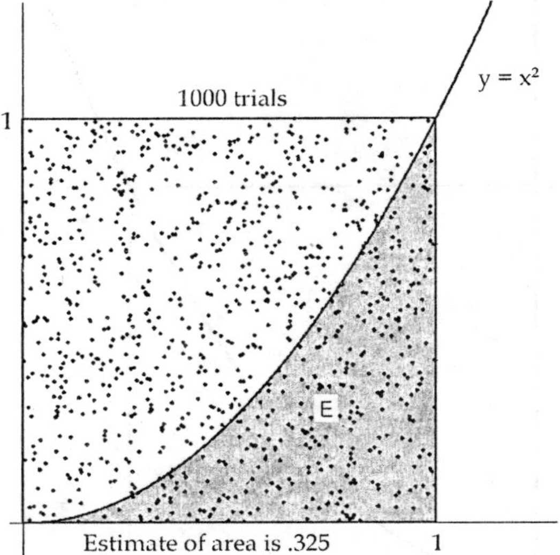

Figure: Computing the area by simulation.

Buffon's Needle

Example: Suppose that we take a card table and draw across the top surface a set of parallel lines a unit distance apart. We then drop a common needle of unit length at random on this surface and observe whether or not the needle lies across one of the lines.

We can describe the possible outcomes of this experiment by coordinates as follows: Let d be the distance from the centre of the needle to the nearest line. Next, let *L* be the line determined by the needle, and define θ as the acute angle that the line *L* makes with the set of parallel lines. (The reader should certainly be wary of this description of the sample space.

We are attempting to coordinatise a set of line segments.) Using this description, we have $0 \leq d \leq 1/2$, and $0 \leq \theta \leq \pi/2$. Moreover, we that the needle lies across the nearest line if and only if the hypotenuse of the triangle is less than half the length of the needle, that is,

$$\frac{d}{\sin\theta} < \frac{1}{2}.$$

Now we assume that when the needle drops, the pair (θ, d) is chosen at random from the rectangle $0 \leq \theta \leq \pi/2$, $0 \leq d \leq 1/2$. We observe whether the needle lies across the nearest line (i.e., whether $d \leq (1/2)\sin\theta$).

The probability of this event *E* is the fraction of the area of the rectangle which lies inside *E*.

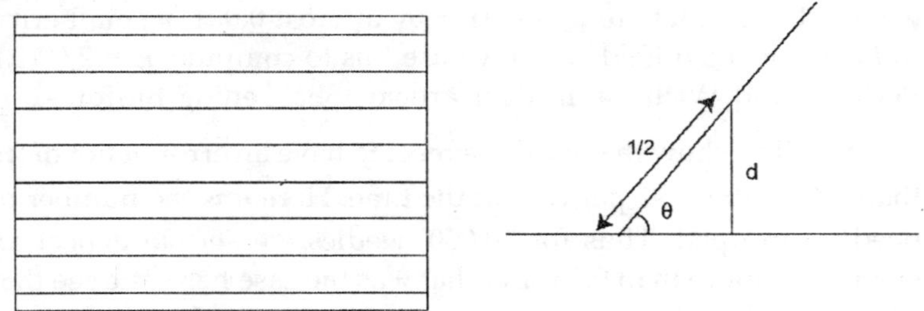

Figure: Buffon's experiment.

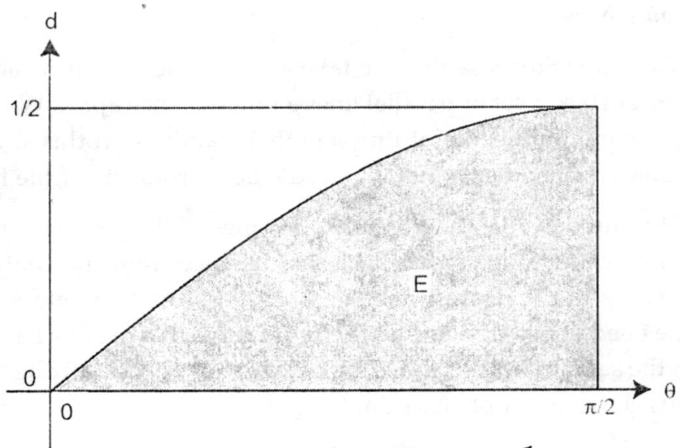

Figure: Set E of pairs (θ, d) with $d < \frac{1}{2}\sin\mu$.

Now the area of the rectangle is $\pi/4$, while the area of E is

$$\text{Area} = \int_0^{\pi/2} \frac{1}{2}\sin\theta d\theta = \frac{1}{2}.$$

Hence, we get

$$P(E) = \frac{1/2}{\pi/4} = \frac{2}{\pi}.$$

The programme *BuffonsNeedle* simulates this experiment. In the following figure, we show the position of every 100th needle in a run of the programme in which 10,000 needles were "dropped." Our final estimate for π is 3.139. While this was within 0.003 of the true value for π we had no right to expect such accuracy. The reason for this is that our simulation estimates $P(E)$. While we can expect this estimate to be in error by at most 0.001, a small error in $P(E)$ gets magnified when we use this to compute $\pi = 2/P(E)$. Perlman and Wichura, in their article "Sharpening Buffon's

Needle," show that we can expect to have an error of not more than $5/\sqrt{n}$ about 95 per cent of the time. Here n is the number of needles dropped. Thus for 10,000 needles we should expect an error of no more than 0.05, and that was the case here. We see that a large number of experiments is necessary to get a decent estimate for π.

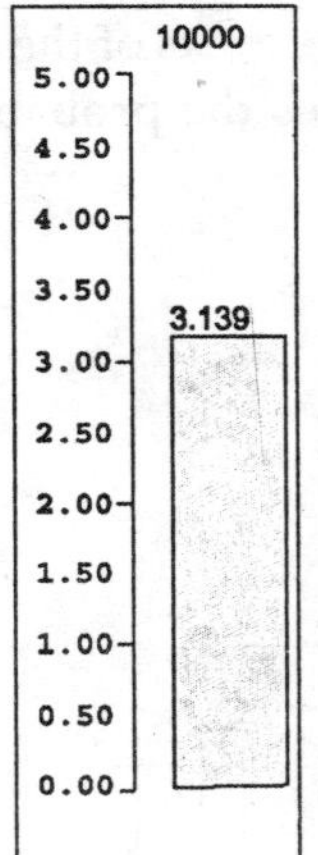

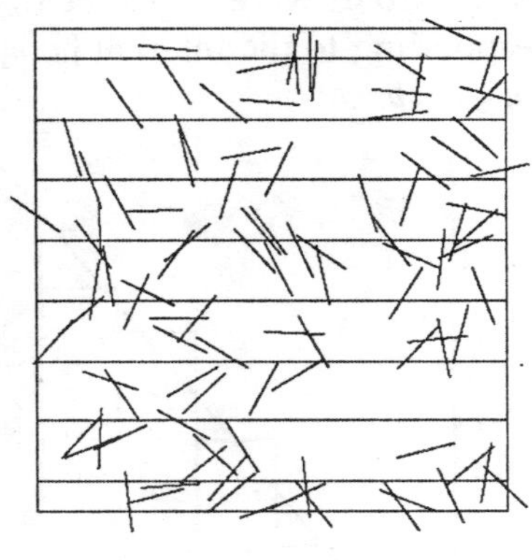

Figure: Simulation of Buffon's needle experiment.

In each of our examples so far, events of the same size are equally likely. Here is an example where they are not. We will see many other such examples later.

Example: Suppose that we choose two random real numbers in [0, 1] and add them together. Let X be the sum. How is X distributed?

To help understand the answer to this question, we can use the programme *Are-abargraph*. This programme produces a bar graph with the property that on each interval, the area, rather than the height, of the bar is equal to the fraction of outcomes that fell in the corresponding interval. We have carried out this experiment 1000 times; the data is shown in the following figure. It appears that the function defined by

$$f(x) = \begin{cases} x, & \text{if } 0 \le x \le 1, \\ 2 - x & \text{if } 1 < x \le 2 \end{cases}$$

fits the data very well. (It is shown in the following figure.) By this we mean that if a and b are any two real numbers between 0 and 2, with $a \le b$, then we can use this function to calculate the probability that $a \le X \le b$. To understand how this calculation might be performed, we again consider the following figure. Because of the way the bars were constructed, the sum of the areas of the bars corresponding to the interval $[a, b]$ approximates the probability that $a \le X \le b$.

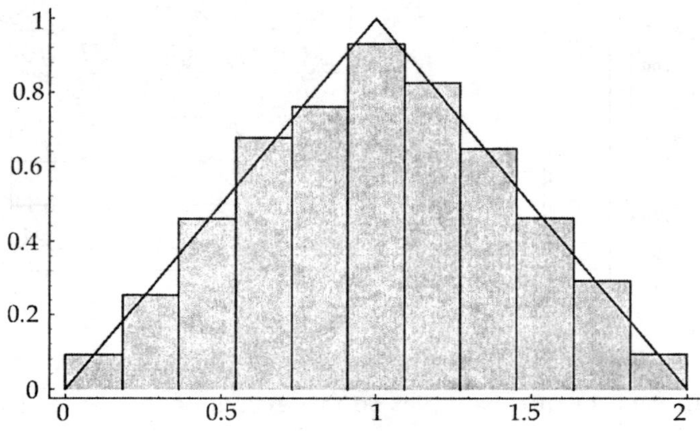

Figure: Sum of two random numbers.

But the sum of the areas of these bars also approximates the integral

$$\int_a^b f(x)\,dx\,.$$

This suggests that for an experiment with a continuum of possible outcomes, if we find a function with the above property, then we will be able to use it to calculate probabilities.

Example: Suppose that we choose 100 random numbers in [0, 1], and let X represent their sum. How is X distributed? We have carried out this experiment 10000 times; the results are shown in the following figure.

It is not so clear what function fits the bars in this case. It turns out that the type of function which does the job is called a normal density function. This type of function is sometimes referred to as a "bell-shaped" curve. It is among the most important functions in the subject of probability.

Our last example explores the fundamental question of how probabilities are assigned.

Bertrand's Paradox

Example: A chord of a circle is a line segment both of whose endpoints lie on probability that its length exceeds $\sqrt{3}$?

Our answer will depend on what we mean by random, which will depend, in turn, on what we choose for coordinates. The sample space Ω is the set of all possible chords in the circle. To find coordinates for these chords, we first introduce a rectangular coordinate system with origin at the centre of the circle.

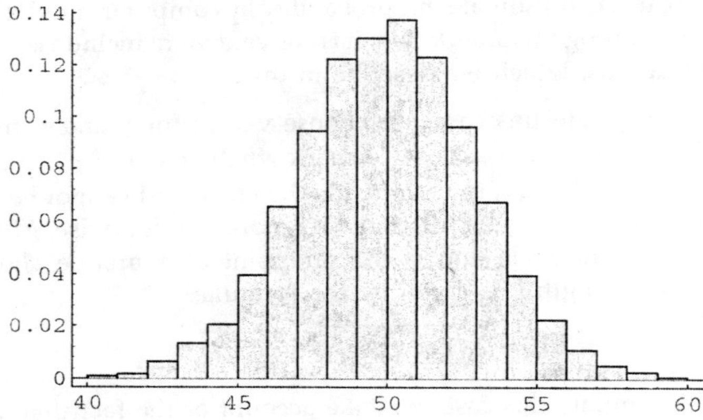

Figure: Sum of 100 random numbers.

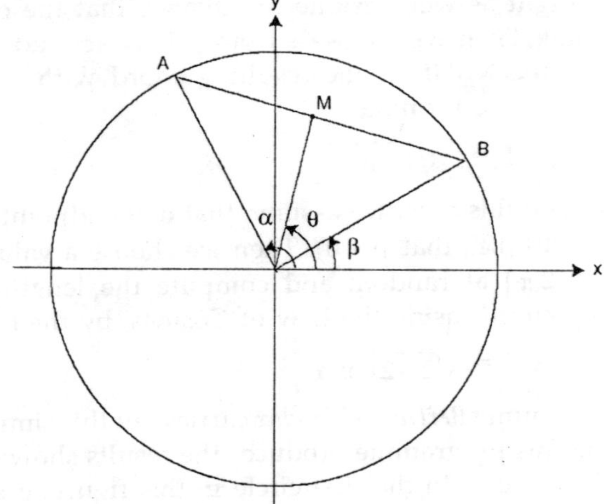

Figure: Random chord.

We note that a chord of a circle is perpendicular to the radial line containing the midpoint of the chord. We can describe each chord by giving:

1. The rectangular coordinates (x, y) of the midpoint M, or
2. The polar coordinates (r, θ) of the midpoint M, or
3. The polar coordinates $(1, \alpha)$ and $(1, \beta)$ of the endpoints A and B.

In each case we shall interpret at random to mean: choose these coordinates at random.

We can easily estimate this probability by computer simulation. In programming this simulation, it is convenient to include certain simplifications, which we describe in turn:

1. To simulate this case, we choose values for x and y from [–1, 1] at random. Then we check whether $x^2 + y^2 \leq 1$. If not, the point $M = (x, y)$ lies outside the circle and cannot be the midpoint of any chord, and we ignore it. Otherwise, M lies inside the circle and is the midpoint of a unique chord, whose length L is given by the formula:

$$L = 2\sqrt{1-(x^2-y^2)}$$

2. To simulate this case, we take account of the fact that any rotation of the circle does not change the length of the chord, so we might as well assume in advance that the chord is horizontal. Then we choose r from [–1, 1] at random, and compute the length of the resulting chord with midpoint (r, $\pi/2$) by the formula:

$$L = 2\sqrt{1-r^2}\,.$$

3. To simulate this case, we assume that one endpoint, say B, lies at (1, 0) (i.e., that $\beta = 0$). Then we choose a value for α from $[0, 2\pi]$ at random and compute the length of the resulting chord, using the Law of Cosines, by the formula:

$$L = \sqrt{2-2\cos\alpha}\,\cdot$$

The programme *BertrandsParadox* carries out this simulation. Running this programme produces the results shown in the following figure. In the first circle in this figure, a smaller circle has been drawn. Those chords which intersect this smaller circle have length at least $\sqrt{3}$. In the second circle in the figure, the vertical line intersects all chords of length at least $\sqrt{3}$. In the third circle, again the vertical line intersects all chords of length at least $\sqrt{3}$.

In each case we run the experiment a large number of times and record the fraction of these lengths that exceed $\sqrt{3}$. We have printed the results of every 100th trial up to 10,000 trials.

It is interesting to observe that these fractions are not the same in the three cases; they depend on our choice of coordinates. This phenomenon was first observed by Bertrand, and is now known as Bertrand's paradox. It is actually not a paradox at all; it is merely a reflection of the fact that different choices of coordinates will lead to different assignments of probabilities. Which assignment is "correct" depends on what application or interpretation of the model one has in mind.

One can imagine a real experiment involving throwing long straws at a circle drawn on a card table. A "correct" assignment of coordinates should not depend on where the circle lies on the card table, or where the card table sits in the room. Jaynes has shown that the only assignment which meets this requirement is (2). In this sense, the assignment (2) is the natural, or "correct" one.

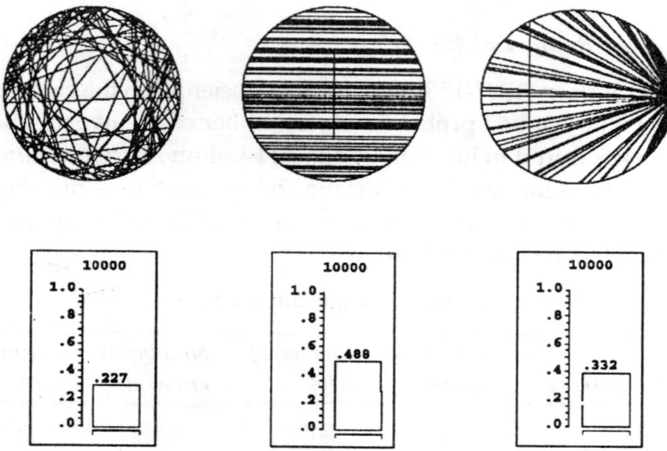

Figure: Bertrand's paradox.

We can easily see in each case what the true probabilities are if we note that $\sqrt{3}$ is the length of the side of an inscribed equilateral triangle. Hence, a chord has length $L > \sqrt{3}$ if its midpoint has distance $d < 1/2$ from the origin.

The following calculations determine the probability that $L > \sqrt{3}$ in each of the three cases.

1. $L > \sqrt{3}$ if (x, y) lies inside a circle of radius $1/2$, which occurs with probability

$$p = \frac{\pi(1/2)^2}{\pi(1)^2} = \frac{1}{4}.$$

2. $L > \sqrt{3}$ if $|r| < 1/2$, which occurs with probability

$$\frac{1/2-(-1/2)}{1-(-1)} = \frac{1}{2}.$$

3. $L > \sqrt{3}$ if $2\pi/3 < \alpha < 4\pi/3$, which occurs with probability

$$\frac{4\pi/3-2\pi/3}{2\pi-0} = \frac{1}{3}.$$

We see that our simulations agree quite well with these theoretical values.

Historical Remarks

G. L. Buffon (1707-1788) was a natural scientist in the eighteenth century who applied probability to a number of his investigations. His work is found in his monumental 44-volume *Histoire Naturelle* and its supplements. For example, he presented a number of mortality tables and used them to compute, for each age group, the expected remaining lifetime.

Table: Buffon needle experiments to estimate π.

Experimenter	*Length of needle*	*Number of casts*	*Number of crossings*	*Estimate for π*
Wolf, 1850	.8	5000	2532	3.1596
Smith, 1855	.6	3204	1218.5	3.1553

De Morgan, c.1860	1.0	600	382.5	3.137
Fox, 1864	.75	1030	489	3.1595
Lazzerini, 1901	.83	3408	1808	3.1415929
Reina, 1925	.5419	2520	869	3.1795

From his table he observed: the expected remaining lifetime of an infant of one year is 33 years, while that of a man of 21 years is also approximately 33 years. Thus, a father who is not yet 21 can hope to live longer than his one year old son, but if the father is 40, the odds are already 3 to 2 that his son will outlive him.

Buffon wanted to show that not all probability calculations rely only on algebra, but that some rely on geometrical calculations. One such problem was his famous "needle problem". In his original formulation, Buffon describes a game in which two gamblers drop a loaf of French bread on a wide-board floor and bet on whether or not the loaf falls across a crack in the floor. Buffon asked: what length L should the bread loaf be, relative to the width W of the floorboards, so that the game is fair. He found the correct answer ($L = (\pi/4)W$) using essentially the methods described in this chapter. He also considered the case of a checkerboard floor, but gave the wrong answer in this case. The correct answer was given later by Laplace.

The literature contains descriptions of a number of experiments that were actually carried out to estimate π by this method of dropping needles. N. T. Gridgeman discusses the experiments shown in the previous table. (The halves for the number of crossing comes from a compromise when it could not be decided if a crossing had actually occurred.) He observes, as we have, that 10,000 casts could do no more than establish the first decimal place of π with reasonable confidence. Gridgeman points out that, although none of the experiments used even 10,000 casts, they are surprisingly good, and in some cases, too good. The fact that the number of casts is not always a round number would suggest that the authors might have resorted to clever stopping to get a good answer. Gridgeman comments that Lazzerini's estimate turned out to agree with a well-known approximation to π, 355/113 =

3:1415929, discovered by the fifth-century Chinese mathematician, Tsu Ch'ungchih. Gridgeman says that he did not have Lazzerini's original report, and while waiting for it (knowing only the needle crossed a line 1808 times in 3408 casts) deduced that the length of the needle must have been 5/6. He calculated this from Buffon's formula, assuming

$$L = \frac{\pi P(E)}{2} = \frac{1}{2}\left(\frac{355}{113}\right)\left(\frac{1808}{3408}\right) = \frac{5}{6} = .8333\cdot$$

Even with careful planning one would have to be extremely lucky to be able to stop so cleverly.

The second author likes to trace his interest in probability theory to the Chicago World's Fair of 1933 where he observed a mechanical device dropping needles and displaying the ever-changing estimates for the value of π. (The first author likes to trace his interest in probability theory to the second author.)

Exercises

1. In the spinner problem divide the unit circumference into three arcs of length 1/2, 1/3, and 1/6. Write a programme to simulate the spinner experiment 1000 times and print out what fraction of the outcomes fall in each of the three arcs. Now plot a bar graph whose bars have width 1/2, 1/3, and 1/6, and areas equal to the corresponding fractions as determined by your simulation. Show that the heights of the bars are all nearly the same.
2. Do the same as in Exercise 1, but divide the unit circumference into five arcs of length 1/3, 1/4, 1/5, 1/6, and 1/20.
3. Alter the programme *MonteCarlo* to estimate the area of the circle of radius 1/2 with centre at (1/2, 1/2) inside the unit square by choosing 1000 points at random. Compare your results with the true value of π =4. Use your results to estimate the value of π. How accurate is your estimate?
4. Alter the programme *MonteCarlo* to estimate the area under the graph of $y = \sin \pi x$ inside the unit square by choosing

10,000 points at random. Now calculate the true value of this area and use your results to estimate the value of π. How accurate is your estimate?

5. Alter the programme *MonteCarlo* to estimate the area under the graph of $y = 1/(x+1)$ in the unit square in the same way as in previous exercise. Calculate the true value of this area and use your simulation results to estimate the value of log 2. How accurate is your estimate?

6. To simulate the Buffon's needle problem we choose independently the distance d and the angle θ at random, with $0 \le d \le 1/2$ and $0 \le \theta \le \pi/2$, and check whether $d \le (1/2)\sin\theta$. Doing this a large number of times, we estimate π as $2/a$, where a is the fraction of the times that $d \le (1/2)\sin\theta$. Write a programme to estimate π by this method. Run your programme several times for each of 100, 1000, and 10,000 experiments. Does the accuracy of the experimental approximation for π improve as the number of experiments increases?

7. For Buffon's needle problem, Laplace considered a grid with horizontal and vertical lines one unit apart. He showed that the probability that a needle of length $L \le 1$ crosses at least one line is

$$p = \frac{4L - L^2}{\pi}.$$

To simulate this experiment we choose at random an angle θ between 0 and $\pi/2$ and independently two numbers d_1 and d_2 between 0 and $L/2$. (The two numbers represent the distance from the centre of the needle to the nearest horizontal and vertical line.) The needle crosses a line if either $d_1 \le (L/2)\sin\theta$ or $d_2 \le (L/2)\cos\theta$. We do this a large number of times and estimate π as

$$\bar{\pi} = \frac{4L - L^2}{a},$$

where a is the proportion of times that the needle crosses at least one line. Write a programme to estimate π by this method, run your programme for 100, 1000, and 10,000 experiments, and compare your results with Buffon's method described in Exercise 6. (Take L = 1.)

8. A long needle of length L much bigger than 1 is dropped on a grid with horizontal and vertical lines one unit apart. We will see that the average number a of lines crossed is approximately

$$a = \frac{4L}{\pi}.$$

To estimate π by simulation, pick an angle θ at random between 0 and $\pi/2$ and compute $L\sin\theta + L\cos\theta$. This may be used for the number of lines crossed. Repeat this many times and estimate π by

$$\overline{\pi} = \frac{4L}{a},$$

where a is the average number of lines crossed per experiment. Write a programme to simulate this experiment and run your programme for the number of experiments equal to 100, 1000, and 10,000. Compare your results with the methods of Laplace or Buffon for the same number of experiments. (Use L = 100.)

The following exercises involve experiments in which not all outcomes are equally likely.

9. A large number of waiting time problems have an exponential distribution of outcomes. Such outcomes are simulated by computing $(-1/\lambda)$ log(rnd), where $\lambda > 0$. For waiting times produced in this way, the average waiting time is $1/\lambda$ For example, the times spent waiting for a car to pass on a highway, or the times between emissions of particles from a radioactive source, are simulated by a sequence of random numbers, each of which is chosen by computing $(-1/\lambda)$ log(rnd), where $1/\lambda$ is the average time between cars or emissions. Write a programme to simulate the times between cars when the average time between cars is 30 seconds. Have your programme compute an area bar graph for these times by breaking the time interval from 0 to 120 into 24

subintervals. On the same pair of axes, plot the function $f(x) = (1/30)e^{-(1/30)x}$. Does the function fit the bar graph well?

10. In previous exercise, the distribution came "out of a hat." In this problem, we will again consider an experiment whose outcomes are not equally likely. We will determine a function $f(x)$ which can be used to determine the probability of certain events. Let T be the right triangle in the plane with vertices at the points (0, 0); (1, 0); and (0, 1). The experiment consists of picking a point at random in the interior of T, and recording only the x-coordinate of the point. Thus, the sample space is the set [0, 1], but the outcomes do not seem to be equally likely. We can simulate this experiment by asking a computer to return two random real numbers in [0, 1], and recording the first of these two numbers if their sum is less than 1. Write this programme and run it for 10,000 trials. Then make a bar graph of the result, breaking the interval [0, 1] into 10 intervals. Compare the bar graph with the function $f(x) = 2 - 2x$. Now show that there is a constant c such that the height of T at the x-coordinate value x is c times $f(x)$ for every x in [0, 1]. Finally, show that

$$\int_0^1 f(x)dx = 1.$$

How might one use the function $f(x)$ to determine the probability that the outcome is between .2 and .5?

11. Here is another way to pick a chord at random on the circle of unit radius. Imagine that we have a card table whose sides are of length 100. We place coordinate axes on the table in such a way that each side of the table is parallel to one of the axes, and so that the centre of the table is the origin. We now place a circle of unit radius on the table so that the centre of the circle is the origin. Now pick out a point (x_0, y_0) at random in the square, and an angle θ at random in the interval $(-\pi/2, \pi/2)$. Let $m = \tan\theta$. Then the equation of the line passing through (x_0, y_0) with slope m is

$$y = y_0 + m(x - x_0),$$

and the distance of this line from the centre of the circle (i.e., the origin) is

$$d = \left| \frac{y_0 - mx_0}{\sqrt{m^2 + 1}} \right|.$$

We can use this distance formula to check whether the line intersects the circle (i.e., whether d < 1). If so, we consider the resulting chord a random chord.

This describes an experiment of dropping a long straw at random on a table on which a circle is drawn.

Write a programme to simulate this experiment 10000 times and estimate the probability that the length of the chord is greater than $\sqrt{3}$.

Continuous Density Functions

We have seen how to simulate experiments with a whole continuum of possible outcomes and have gained some experience in thinking about such experiments. Now we turn to the general problem of assigning probabilities to the outcomes and events in such experiments. We shall restrict our attention here to those experiments whose sample space can be taken as a suitably chosen subset of the line, the plane, or some other Euclidean space. We begin with some simple examples.

Spinners

Example: We would like to construct a probability model in which each outcome is equally likely to occur. We saw that in such a model, it is necessary to assign the probability 0 to each outcome. This does not at all mean that the probability of every event must be zero. On the contrary, if we let the random variable X denote the outcome, then the probability

$$P\ (0 \leq X \leq 1)$$

that the head of the spinner comes to rest somewhere in the circle, should be equal to 1. Also, the probability that it comes to rest in the upper half of the circle should be the same as for the lower half, so that

$$P\left(0 \le X < \frac{1}{2}\right) = P\left(\frac{1}{2} \le X < 1\right) = \frac{1}{2}.$$

More generally, in our model, we would like the equation

$$P(c \le X < d) = d - c$$

to be true for every choice of c and d.

If we let $E = [c, d]$, then we can write the above formula in the form

$$P(E) = \int_E f(x)\,dx,$$

where $f(x)$ is the constant function with value 1. This should remind the reader of the corresponding formula in the discrete case for the probability of an event:

$$P(E) = \sum_{w \in E} m(w).$$

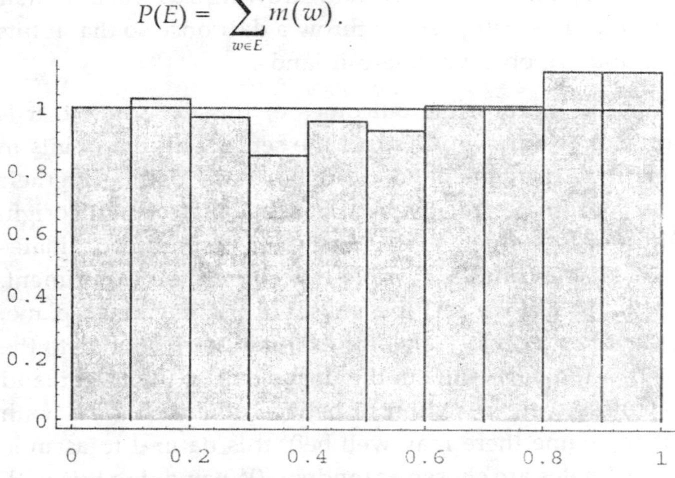

Figure: Spinner experiment.

The difference is that in the continuous case, the quantity being integrated, $f(x)$, is not the probability of the outcome x. (However, if one uses infinitesimals, one can consider $f(x)dx$ as the probability of the outcome x.)

In the continuous case, we will use the following convention. If the set of outcomes is a set of real numbers, then the individual outcomes will be referred to by small Roman letters such as x. If the set of outcomes is a subset of R^2, then the individual outcomes will be denoted by (x, y). In either case, it may be more convenient to refer to an individual outcome by using w.

The previous figure shows the results of 1000 spins of the spinner. The function $f(x)$ is also shown in the figure. The reader will note that the area under $f(x)$ and above a given interval is approximately equal to the fraction of outcomes that fell in that interval. The function $f(x)$ is called the density function of the random variable X. The fact that the area under $f(x)$ and above an interval corresponds to a probability is the defining property of density functions. A precise definition of density functions will be given shortly.

Darts

Example: A game of darts involves throwing a dart at a circular target of unit radius. Suppose we throw a dart once so that it hits the target, and we observe where it lands.

To describe the possible outcomes of this experiment, it is natural to take as our sample space the set Ω of all the points in the target. It is convenient to describe these points by their rectangular coordinates, relative to a coordinate system with origin at the centre of the target, so that each pair (x, y) of coordinates with $x^2 + y^2 \leq 1$ describes a possible outcome of the experiment. Then $\Omega = \{(x, y) : x^2 + y^2 \leq 1\}$ is a subset of the Euclidean plane, and the event $E = \{(x, y) : y > 0\}$, for example, corresponds to the statement that the dart lands in the upper half of the target, and so forth. Unless there is reason to believe otherwise (and with experts at the game there may well be!), it is natural to assume that the coordinates are chosen at random. (When doing this with a computer, each coordinate is chosen uniformly from the interval [-1, 1]. If the resulting point does not lie inside the unit circle, the point is not counted.) Then the arguments used in the preceding example show that the probability of any elementary event, consisting of a single outcome, must be zero, and suggest that the probability of the event that the dart lands in any subset E of the target should be determined by what fraction of the target area lies in E . Thus,

$$P(E) = \frac{\text{area of } E}{\text{area of target}} = \frac{\text{area of } E}{\pi}.$$

This can be written in the form

$$P(E) = \int_E f(x)dx,$$

where $f(x)$ is the constant function with value $1/\pi$. In particular, if $E = \{(x, y) : x^2 + y^2 \leq a^2\}$ is the event that the dart lands within distance $a < 1$ of the centre of the target, then

$$P(E) = \frac{\pi a^2}{\pi} = a^2.$$

For example, the probability that the dart lies within a distance 1/2 of the centre is 1/4.

Example: In the dart game considered above, suppose that, instead of observing where the dart lands, we observe how far it lands from the centre of the target. In this case, we take as our sample space the set Ω of all circles with centres at the centre of the target. It is convenient to describe these circles by their radii, so that each circle is identified by its radius r, $0 \leq r \leq 1$. In this way, we may regard Ω as the subset [0, 1] of the real line.

What probabilities should we assign to the events E of Ω ? If

$$E = \{r : 0 \leq r \leq a\},$$

then E occurs if the dart lands within a distance a of the centre, that is, within the circle of radius a, and we saw in the previous example that under our assumptions the probability of this event is given by

$$P([0, a]) = a^2.$$

More generally, if

$$E = \{r : a \leq r \leq b\},$$

then by our basic assumptions,

$$
\begin{aligned}
P(E) = P([a, b]) &= P([0, b]) - P([0, a]) \\
&= b^2 - a^2 \\
&= (b - a)(b + a) \\
&= 2(b - a)\,\frac{(b + a)}{2}
\end{aligned}
$$

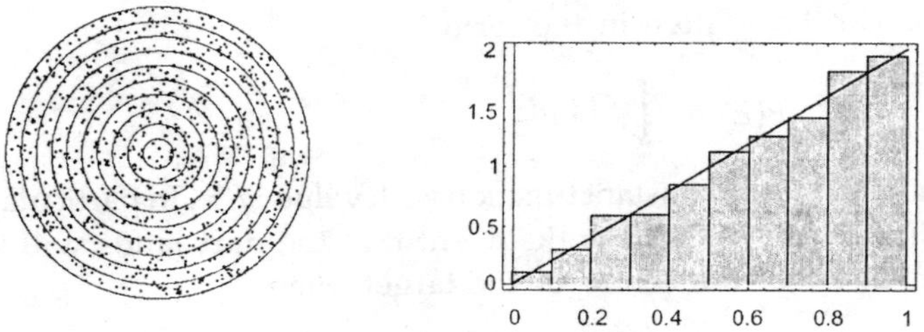

Figure: Distribution of dart distances in 400 throws.

Thus, $P(E)$ = 2(length of E)(midpoint of E). Here we see that the probability assigned to the interval E depends not only on its length but also on its midpoint (i.e., not only on how long it is, but also on where it is). Roughly speaking, in this experiment, events of the form $E = [a, b]$ are more likely if they are near the rim of the target and less likely if they are near the centre. (A common experience for beginners! The conclusion might well be different if the beginner is replaced by an expert.)

Again we can simulate this by computer. We divide the target area into ten concentric regions of equal thickness.

The computer programme *Darts* throws n darts and records what fraction of the total falls in each of these concentric regions. The programme *Areabargraph* then plots a bar graph with the area of the ith bar equal to the fraction of the total falling in the ith region. Running the programme for 1000 darts resulted in the bar graph of the previous figure.

Note that here the heights of the bars are not all equal, but grow approximately linearly with r. In fact, the linear function $y = 2r$ appears to fit our bar graph quite well. This suggests that the probability that the dart falls within a distance a of the centre should be given by the area under the graph of the function $y = 2r$ between 0 and a. This area is a^2, which agrees with the probability we have assigned above to this event.

Sample Space Coordinates

These examples suggest that for continuous experiments of this sort we should assign probabilities for the outcomes to fall in a given interval by means of the area under a suitable function.

More generally, we suppose that suitable coordinates can be introduced into the sample space Ω, so that we can regard Ω as a subset of R^n. We call such a sample space a continuous sample space. We let X be a random variable which represents the outcome of the experiment. Such a random variable is called a continuous random variable. We then define a density function for X as follows.

Density Functions of Continuous Random Variables

Definition: Let X be a continuous real-valued random variable. A density function for X is a real-valued function f which satisfies

$$P(a \le X \le b) = \int_b^a f(x)\,dx$$

for all a, b $\in$ R.

We note that it is not the case that all continuous real-valued random variables possess density functions. However, in this book, we will only consider continuous random variables for which density functions exist.

In terms of the density $f(x)$, if E is a subset of R, then

$$P(X \in E) = \int_E f(x)\,dx.$$

The notation here assumes that E is a subset of R for which $\int_E f(x)\,dx$ make sense.

Example: In the spinner experiment, we choose for our set of outcomes the interval $0 \le x \le 1$, and for our density function

$$f(x) = \begin{cases} 1, & \text{if } 0 \le x < 1, \\ 0, & \text{otherwise.} \end{cases}$$

If E is the event that the head of the spinner falls in the upper half of the circle, then $E = \{x : 0 \le x \le 1/2\}$, and so

$$P(E) = \int_0^{1/2} 1dx \cdot = \frac{1}{2}$$

More generally, if E is the event that the head falls in the interval $[a, b]$, then

$$P(E) = \int_a^b 1dx = b - a \cdot$$

Example: In the first dart game experiment, we choose for our sample space a disc of unit radius in the plane and for our density function the function

$$f(x, y) = \begin{cases} 1/\pi, & \text{if } x^2 + y^2 \le 1, \\ 0, & \text{otherwise.} \end{cases}$$

The probability that the dart lands inside the subset E is then given by

$$P(E) = \iint_E \frac{1}{\pi} dxdy$$

$$= \frac{1}{\pi} \cdot \text{(area of } E\text{)}.$$

In these two examples, the density function is constant and does not depend on the particular outcome. It is often the case that experiments in which the coordinates are chosen at random can be described by constant density functions, we call such density functions uniform or equiprobable. Not all experiments are of this type, however.

Example: In the second dart game experiment, we choose for our sample space the unit interval on the real line and for our density the function

$$f(r) = \begin{cases} 2r, & \text{if } 0 < r < 1, \\ 0, & \text{otherwise.} \end{cases}$$

Then the probability that the dart lands at distance r, $a \leq r \leq b$, from the centre of the target is given by

$$P([a, b]) = \int_a^b 2r\,dr$$

$$= b^2 - a^2.$$

Here again, since the density is small when r is near 0 and large when r is near 1, we see that in this experiment the dart is more likely to land near the rim of the target than near the centre. The heights of the bars approximate the density function, while the areas of the bars approximate the probabilities of the subintervals.

We see in this example that, unlike the case of discrete sample spaces, the value $f(x)$ of the density function for the outcome x is not the probability of x occurring (we have seen that this probability is always 0) and in general $f(x)$ is not a probability at all. In this example, if we take $\lambda = 2$ then $f(3/4) = 3/2$, which being bigger than 1, cannot be a probability.

Nevertheless, the density function f does contain all the probability information about the experiment, since the probabilities of all events can be derived from it. In particular, the probability that the outcome of the experiment falls in an interval $[a, b]$ is given by

$$P([a, b]) = \int_a^b f(x)\,dx,$$

that is, by the *area* under the graph of the density function in the interval $[a, b]$. Thus, there is a close connection here between probabilities and areas. We have been guided by this close connection in making up our bar graphs; each bar is chosen so that its area, and not its height, represents the relative frequency of occurrence, and hence estimates the probability of the outcome falling in the associated interval.

In the language of the calculus, we can say that the probability of occurrence of an event of the form $[x, x + dx]$, where dx is small, is approximately given by

$$P([x, x + dx]) \approx f(x)dx,$$

that is, by the area of the rectangle under the graph of f. Note that as $dx \to 0$, this probability $\to 0$, so that the probability $P(\{x\})$ of a single point is again 0.

Cumulative Distribution Functions of Continuous Random Variables

We have seen that density functions are useful when considering continuous random variables. There is another kind of function, closely related to these density functions, which is also of great importance. These functions are called cumulative distribution functions.

Definition: Let X be a continuous real-valued random variable. Then the cumulative distribution function of X is defined by the equation

$$F_X(x) = P(X \leq x).$$

If X is a continuous real-valued random variable which possesses a density function, then it also has a cumulative distribution function, and the following theorem shows that the two functions are related in a very nice way.

Theorem: Let X be a continuous real-valued random variable with density function $f(x)$. Then the function defined by

$$F(x) = \int_{-\infty}^{x} f(t)\,dt$$

is the cumulative distribution function of X. Furthermore, we have

$$\frac{d}{dx}F(x) = f(x).$$

Proof. By definition,

$$F(x) = P(X \leq x).$$

Let $E = (-\infty, x]$. Then

$$P(X \leq x) = P(X \in E),$$

which equals

$$\int_{-\infty}^{x} f(t)\,dt\,.$$

Applying the Fundamental Theorem of Calculus to the first equation in the statement of the theorem yields the second statement.

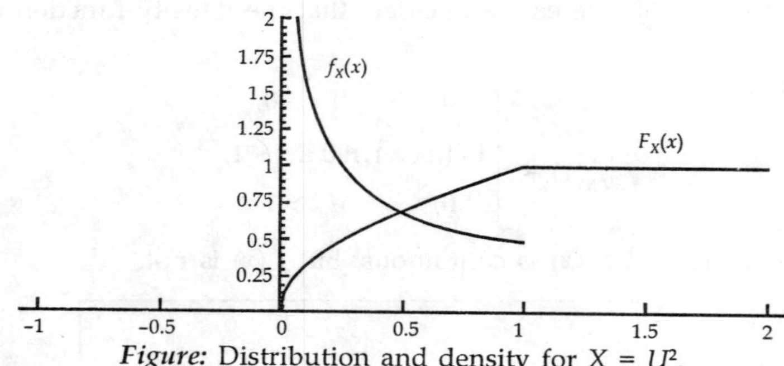

Figure: Distribution and density for $X = U^2$.

In many experiments, the density function of the relevant random variable is easy to write down. However, it is quite often the case that the cumulative distribution function is easier to obtain than the density function. (Of course, once we have the cumulative distribution function, the density function can easily be obtained by differentiation, as the above theorem shows.) We now give some examples which exhibit this phenomenon.

Example: A real number is chosen at random from [0, 1] with uniform probability, and then this number is squared. Let X represent the result. What is the cumulative distribution function of X? What is the density of X?

We begin by letting U represent the chosen real number. Then $X = U^2$. If $0 \le x \le 1$, then we have

$$\begin{aligned} F_X(x) &= P(X \le x) \\ &= P(U^2 \le x) \\ &= P(U \le \sqrt{x}) \\ &= \sqrt{x} \end{aligned}$$

It is clear that X always takes on a value between 0 and 1, so the cumulative distribution function of X is given by

$$F_X(x) = \begin{cases} 0, & \text{if } x \le 0, \\ \sqrt{x}, & \text{if } 0 \le x \le 1, \\ 1, & \text{if } x \ge 1. \end{cases}$$

From this we easily calculate that the density function of X is

$$f_X(x) = \begin{cases} 0, & \text{if } x \le 0, \\ 1/\left(2\sqrt{x}\right), & \text{if } 0 \le x \le 1, \\ 0, & \text{if } x > 1. \end{cases}$$

Note that $F_X(x)$ is continuous, but $f_X(x)$ is not.

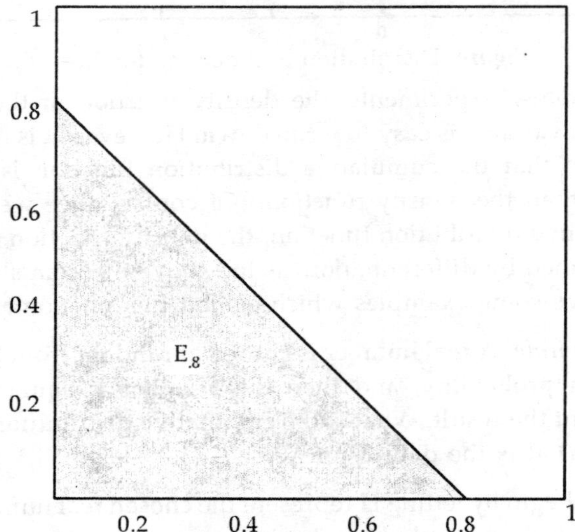

Figure: Calculation of distribution function for the following example.

When referring to a continuous random variable X (say with a uniform density function), it is customary to say that "X is uniformly distributed on the interval $[a, b]$." It is also customary to refer to the cumulative distribution function of X as the distribution function of X. Thus, the word "distribution" is being used in several different ways in the subject of probability. (Recall that it also has a meaning when discussing discrete random variables.)

When referring to the cumulative distribution function of a continuous random variable X, we will always use the word "cumulative" as a modifier, unless the use of another modifier, such as "normal" or "exponential," makes it clear. Since the phrase "uniformly densitied on the interval $[a, b]$" is not acceptable English, we will have to say "uniformly distributed" instead.

Example: Let the random variables X and Y denote the two chosen real numbers. Define $Z = X + Y$. We will now derive expressions for the cumulative distribution function and the density function of Z.

Here we take for our sample space Ω the unit square in R^2 with uniform density. A point $w \in \Omega$ then consists of a pair (x, y) of numbers chosen at random. Then $0 \leq Z \leq 2$. Let E_z denote the event that $Z \leq z$. In the previous figure, we show the set $E_{.8}$. The event E_z, for any z between 0 and 1, looks very similar to the shaded set in the figure. For $1 < z \leq 2$, the set E_z looks like the unit square with a triangle removed from the upper right-hand corner. We can now calculate the probability distribution F_Z of Z; it is given by

$$F_Z(z) = P(Z \leq z)$$

$$= \text{Area of } E_z$$

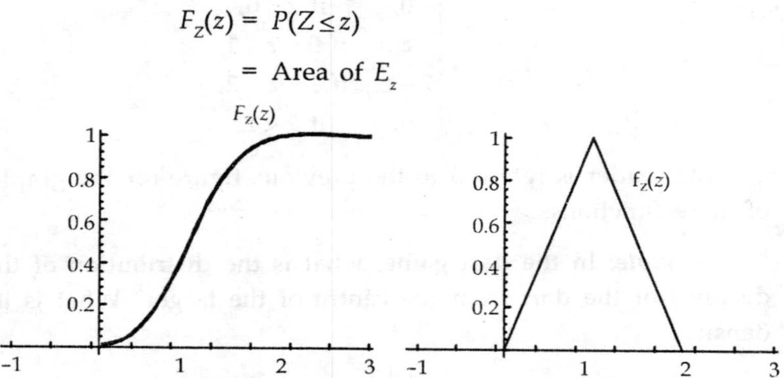

Figure: Distribution and density functions for the previous example.

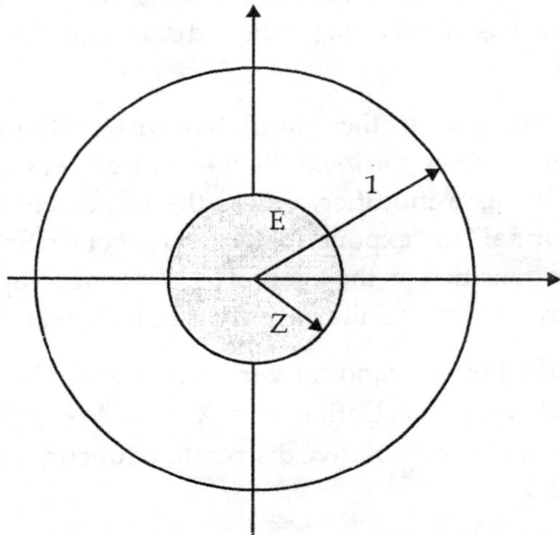

Figure: Calculation of F_z for the following example.

$$\begin{cases} 0, & \text{if } z < 0, \\ (1/2)z^2, & \text{if } 0 \le z \le 1, \\ 1-(1/2)(2-z)^2 & \text{if } 1 \le z \le 2, \\ 1, & \text{if } 2 < z. \end{cases}$$

The density function is obtained by differentiating this function:

$$fz(z) = \begin{cases} 0, & \text{if } z < 0, \\ z, & \text{if } 0 \le z \le 1, \\ 2-z, & \text{if } 1 \le z \le 2, \\ 0, & \text{if } 2 < z. \end{cases}$$

The ıeader is referred to the previous figure for the graphs of these functions.

Example: In the dart game, what is the distribution of the distance of the dart from the centre of the target? What is its density?

Here, as before, our sample space Ω is the unit disk in R^2, with coordinates (X, Y). Let $Z = \sqrt{X^2 + Y^2}$ represent the distance from the centre of the target. Let

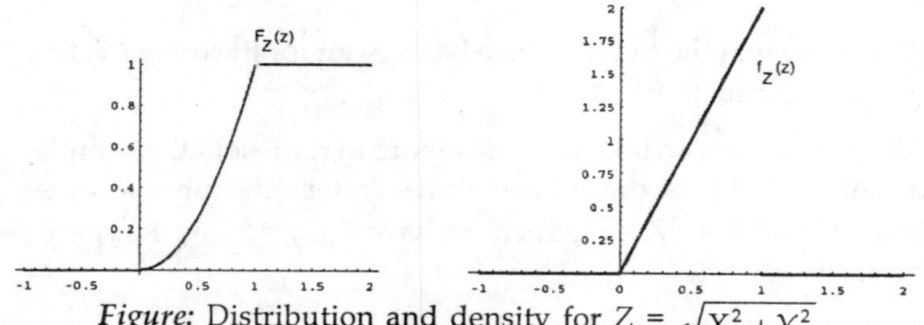

Figure: Distribution and density for $Z = \sqrt{X^2 + Y^2}$.

E be the event $\{Z \le z\}$. Then the distribution function F_Z of Z is given by

$$F_Z(z) = P(Z \le z)$$

$$= \frac{\text{Area of E}}{\text{Area of target}}.$$

Thus, we easily compute that

$$F_Z(z) = \begin{cases} 0, & \text{if } z \le 0, \\ z^2, & \text{if } 0 \le z \le 1, \\ 1, & \text{if } z > 1. \end{cases}$$

The density $f_Z(z)$ is given again by the derivative of $F_Z(z)$:

$$f_Z(z) = \begin{cases} 0, & \text{if } z \le 0, \\ 2z, & \text{if } 0 \le z \le 1, \\ 0, & \text{if } z > 1. \end{cases}$$

The reader is referred to the previous figure for the graphs of these functions.

We can verify this result by simulation, as follows: We choose values for X and Y at random from $[0, 1]$ with uniform distribution, calculate $Z = \sqrt{X^2 + Y^2}$, check whether $0 \le Z \le 1$, and present the results in a bar graph.

Example: Suppose Mr. and Mrs. Lockhorn agree to meet at the Hanover Inn between 5:00 and 6:00 pm on Tuesday. Suppose each arrives at a time between 5:00 and 6:00 chosen at random with uniform probability. What is the distribution function for the length of time that the first to arrive has to wait for the other? What is the density function?

Here again we can take the unit square to represent the sample space, and (*X*, *Y*) as the arrival times (after 5:00 pm) for the Lockhorns. Let $Z = |X-Y|$. Then we have $F_X(x) = x$ and $F_Y(y) = y$. Moreover,

$$\begin{aligned} F_Z(z) &= P(Z \le z) \\ &= P(|X-Y| \le z) \\ &= \text{Area of E.} \end{aligned}$$

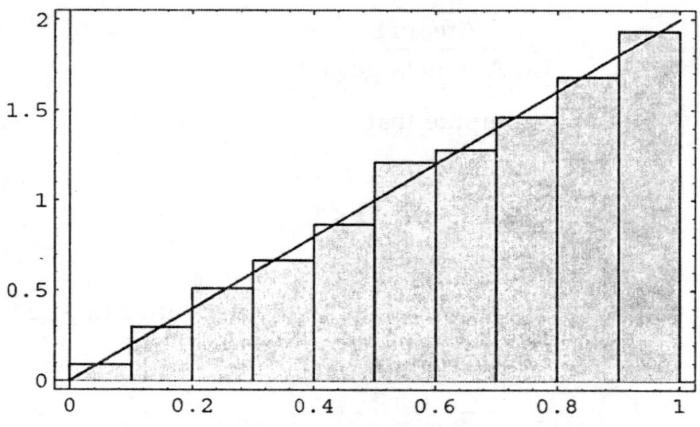

Figure: Simulation results

Thus, we have

$$F_Z(z) = \begin{cases} 0, & \text{if } z \le 0, \\ 1-(1-z)^2, & \text{if } 0 \le z \le 1, \\ 1, & \text{if } z > 1. \end{cases}$$

The density $f_Z(z)$ is again obtained by differentiation:

$$F_Z(z) = \begin{cases} 0, & \text{if } z \le 0, \\ 2(1-z), & \text{if } 0 \le z \le 1, \\ 0, & \text{if } z > 1. \end{cases}$$

Example: There are many occasions where we observe a sequence of occurrences which occur at "random" times. For example, we might be observing emissions of a radioactive isotope, or cars passing a milepost on a highway, or light bulbs burning out. In such cases, we might define a random variable *X* to denote the time between successive occurrences. Clearly, *X* is a continuous random variable whose range consists of the non-negative real numbers. It is often the case that we can model *X* by using the exponential density. This density is given by the formula

$$f(t) = \begin{cases} \lambda e^{-\lambda t}, & \text{if } t \geq 0, \\ 0, & \text{if } t < 0. \end{cases}$$

The number λ is a non-negative real number, and represents the reciprocal of the average value of *X*. Thus, if the average time between occurrences is 30 minutes, then $\lambda = 1/30$. A graph of this density function with $\lambda = 1/30$ is shown in the following figure. One can see from the figure that even though the average value is 30, occasionally much larger values are taken on by *X*.

Suppose that we have bought a computer that contains a Warp 9 hard drive. The salesperson says that the average time between breakdowns of this type of hard drive is 30 months.

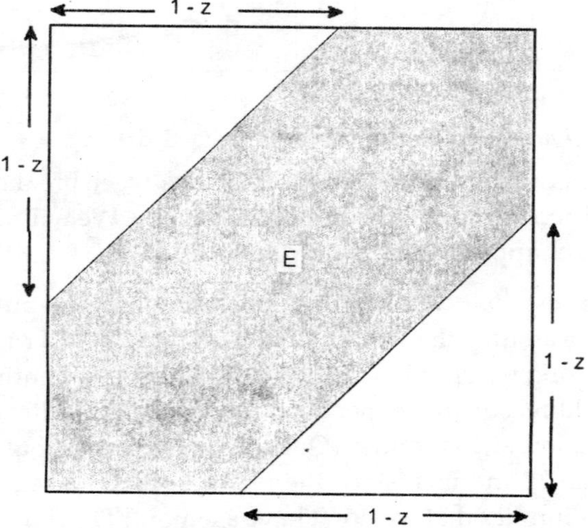

Figure: Calculation of F_Z.

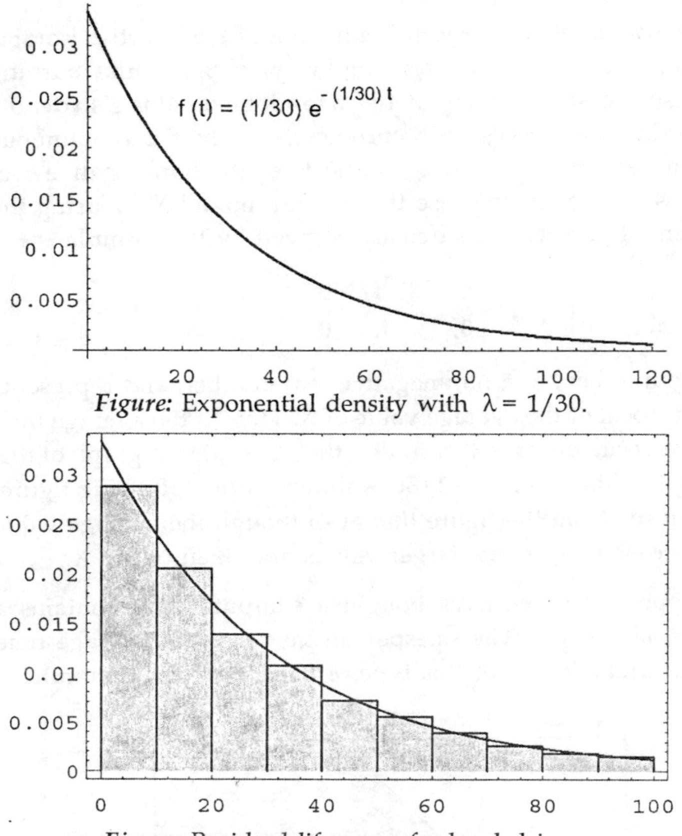

Figure: Exponential density with $\lambda = 1/30$.

Figure: Residual lifespan of a hard drive.

It is often assumed that the length of time between breakdowns is distributed according to the exponential density. We will assume that this model applies here, with $\lambda = 1/30$.

Now suppose that we have been operating our computer for 15 months. We assume that the original hard drive is still running. We ask how long we should expect the hard drive to continue to run. One could reasonably expect that the hard drive will run, on the average, another 15 months. (One might also guess that it will run more than 15 months, since the fact that it has already run for 15 months implies that we don't have a lemon.) The time which we have to wait is a new random variable, which we will call Y.

Obviously, $Y = X - 15$. We can write a computer programme to produce a sequence of simulated Y-values. To do this, we first produce a sequence of X's, and discard those values which are less than or equal to 15 (these values correspond to the cases where the hard drive has quit running before 15 months). To simulate a value of X, we compute the value of the expression

$$\left(-\frac{1}{\lambda}\right)\log(rnd),$$

where *rnd* represents a random real number between 0 and 1. Previous figure shows an area bar graph of 10,000 simulated Y-values.

The average value of Y in this simulation is 29.74, which is closer to the original average lifespan of 30 months than to the value of 15 months which was guessed above. Also, the distribution of Y is seen to be close to the distribution of X. It is in fact the case that X and Y have the same distribution. This property is called the memoryless property, because the amount of time that we have to wait for an occurrence does not depend on how long we have already waited. The only continuous density function with this property is the exponential density.

Assignment of Probabilities

A fundamental question in practice is: How shall we choose the probability density function in describing any given experiment? The answer depends to a great extent on the amount and kind of information available to us about the experiment. In some cases, we can see that the outcomes are equally likely. In some cases, we can see that the experiment resembles another already described by a known density. In some cases, we can run the experiment a large number of times and make a reasonable guess at the density on the basis of the observed distribution of outcomes. In general, the problem of choosing the right density function for a given experiment is a central problem for the experimenter and is not always easy to solve. We shall not examine this question in detail here but instead shall assume that the right density is already known for each of the experiments under study.

The introduction of suitable coordinates to describe a continuous sample space, and a suitable density to describe its probabilities.

Infinite Tree

Example: Consider an experiment in which a fair coin is tossed repeatedly, without stopping. For a coin tossed n times, the natural sample space is a binary tree with n stages. On this evidence we expect that for a coin tossed repeatedly, the natural sample space is a binary tree with an infinite number of stages, as indicated in the figure given below.

It is surprising to learn that, although the n-stage tree is obviously a finite sample space, the unlimited tree can be described as a continuous sample space. To see how this comes about, let us agree that a typical outcome of the unlimited coin tossing experiment can be described by a sequence of the form w = {H H T H T T H . . .}. If we write 1 for H and 0 for T, then w = {1 1 0 1 0 0 1 . . .}. In this way, each outcome is described by a sequence of 0's and 1's.

Now suppose we think of this sequence of 0's and 1's as the binary expansion of some real number x = .1101001 . . . lying between 0 and 1. (A binary expansion is like a decimal expansion but based on 2 instead of 10.) Then each outcome is described by a value of x, and in this way x becomes a coordinate for the sample space, taking on all real values between 0 and 1. (We note that it is possible for two different sequences to correspond to the same real number; for example, the sequences {T H H H H H . . .} and {H T T T T T . . .} both correspond to the real number 1/2. We will not concern ourselves with this apparent problem here.)

What probabilities should be assigned to the events of this sample space? Consider, for example, the event E consisting of all outcomes for which the first toss comes up heads and the second tails. Every such outcome has the form .10 * * * * . . ., where * can be either 0 or 1. Now if x is our real-valued coordinate, then the value of x for every such outcome must lie between

$1/2 = .10000\ldots$ and $3/4 = .11000\ldots$, and moreover, every value of x between 1/2 and 3/4 has a binary expansion of the

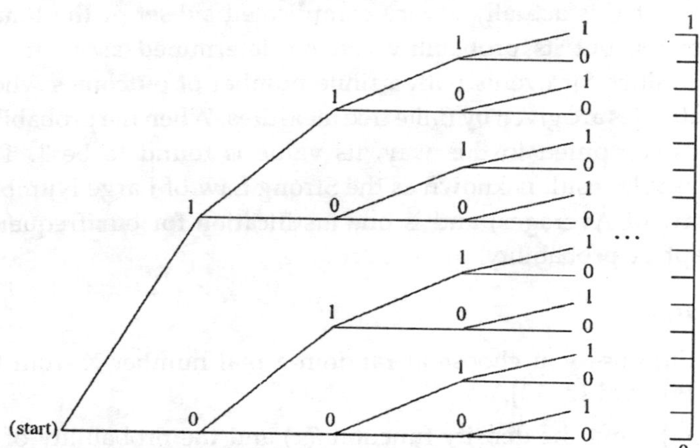

Figure: Tree for infinite number of tosses of a coin.

form $.10****\ldots$. This means that $w \in E$ if and only if $1/2 \leq x \leq 3/4$, and in this way we see that we can describe E by the interval [1/2, 3/4). More generally, every event consisting of outcomes for which the results of the first n tosses are prescribed is described by a binary interval of the form $[k/2^n, (k+1)/2^n)$.

In the experiment involving n tosses, the probability of any one outcome must be exactly $1/2^n$. It follows that in the unlimited toss experiment, the probability of any event consisting of outcomes for which the results of the first n tosses are prescribed must also be $1/2^n$. But $1/2^n$ is exactly the length of the interval of x-values describing E! Thus we see that, just as with the spinner experiment, the probability of an event E is determined by what fraction of the unit interval lies in E.

Consider again the statement: The probability is 1/2 that a fair coin will turn up heads when tossed. We have suggested that one interpretation of this statement is that if we toss the coin indefinitely the proportion of heads will approach 1/2. That is, in our correspondence with binary sequences we expect to get a binary sequence with the proportion of 1's tending to 1/2. The event E of binary sequences for which this is true is a proper subset of the set of all possible binary sequences. It does not contain, for example, the sequence 011011011 . . . (i.e., (011) repeated again and again). The event E is actually a very complicated subset of the binary sequences, but its probability can be determined as a limit of probabilities for events with a finite number of outcomes whose probabilities are given by finite tree measures. When the probability of E is computed in this way, its value is found to be 1. This remarkable result is known as the Strong Law of Large Numbers (or Law of Averages) and is one justification for our frequency concept of probability.

Exercises

1. Suppose you choose at random a real number X from the interval [2, 10].
   (a) Find the density function $f(x)$ and the probability of an event E for this experiment, where E is a subinterval $[a, b]$ of [2, 10].
   (b) From (a), find the probability that $X > 5$, that $5 < X < 7$, and that $X^2 - 12X + 35 > 0$.
2. Suppose you choose a real number X from the interval [2, 10] with a density function of the form
   $$f(x) = C_x,$$
   where C is a constant.
   (a) Find C.
   (b) Find $P(E)$, where $E = [a, b]$ is a subinterval of [2, 10].
   (c) Find $P(X > 5)$, $P(X < 7)$, and $P(X^2 - 12X + 35 > 0)$.
3. Same as Exercise 2, but suppose
   $$f(x) = \frac{C}{x}.$$
4. Suppose you throw a dart at a circular target of radius 10 inches. Assuming that you hit the target and that the coordinates of the outcomes are chosen at random, find the probability that the dart falls
   (a) within 2 inches of the centre.
   (b) within 2 inches of the rim.

(c) within the first quadrant of the target.

(d) within the first quadrant and within 2 inches of the rim.

5. Suppose you are watching a radioactive source that emits particles at a rate described by the exponential density

$$f(t) = \lambda e^{-\lambda t},$$

where $T = 1$, so that the probability $P(0, T)$ that a particle will appear in the next T seconds is $P([0, T]) = \int_0^T \lambda e^{-\lambda t} dt$. Find the probability that a particle (not necessarily the first) will appear

(a) within the next second.

(b) within the next 3 seconds.

(c) between 3 and 4 seconds from now.

(d) after 4 seconds from now.

6. Assume that a new light bulb will burn out after t hours, where t is chosen from $[0, \infty)$ with an exponential density

$$f(t) = \lambda e^{-\lambda t}.$$

In this context, λ is often called the failure rate of the bulb.

(a) Assume that $\lambda = 0.01$, and find the probability that the bulb will not burn out before T hours. This probability is often called the reliability of the bulb.

(b) For what T is the reliability of the bulb = 1/2?

7. Choose a number B at random from the interval [0, 1] with uniform density. Find the probability that

(a) $1/3 < B < 2/3$.

(b) $|B - 1/2| \leq 1/4$.

(c) $B < 1/4$ or $1 - B < 1/4$.

(d) $3B^2 < B$.

8. Choose independently two numbers B and C at random from the interval [0, 1] with uniform density. Note that the point (B, C) is then chosen at random in the unit square. Find the probability that

(a) $B + C < 1/2$.

(b) $BC < 1/2$.

(c) $|B - C| < 1/2$.

(d) $\max\{B, C\} < 1/2$.

(e) $\min\{B, C\} < 1/2$.

(f) $B < 1/2$ and $1 - C < 1/2$.

(g) conditions (c) and (f) both hold.

(h) $B^2 + C^2 \leq 1/2$.

(i) $(B - 1/2)^2 + (C - 1/2)^2 < 1/4$.

9. Suppose that we have a sequence of occurrences. We assume that the time X between occurrences is exponentially distributed with $\lambda = 1/10$, so on the average, there is one occurrence every 10 minutes. You come upon this system at time 100, and wait until the next occurrence. Make a conjecture concerning how long, on the average, you will have to wait. Write a programme to see if your conjecture is right.

10. As in Exercise 9, assume that we have a sequence of occurrences, but now assume that the time X between occurrences is uniformly distributed between 5 and 15. As before, you come upon this system at time 100, and wait until the next occurrence. Make a conjecture concerning how long, on the average, you will have to wait. Write a programme to see if your conjecture is right.

11. For examples such as those in Exercises 9 and 10, it might seem that at leastyou should not have to wait on average more than 10 minutes if the average time between occurrences is 10 minutes. Alas, even this is not true. To see why, consider the following assumption about the times between occurrences. Assume that the time between occurrences is 3 minutes with probability .9 and 73 minutes with probability .1. Show by simulation that the average time between occurrences is 10 minutes, but that if you come upon this system at time 100, your average waiting time is more than 10 minutes.

12. Take a stick of unit length and break it into three pieces, choosing the break points at random. (The break points are assumed to be chosen simultaneously.) What is the probability that the three pieces can be used to form a triangle? *Hint*: The sum of the lengths of any two pieces must exceed the length of the third, so each piece must have length < 1/2.
13. Take a stick of unit length and break it into two pieces, choosing the break point at random. Now break the longer of the two pieces at a random point. What is the probability that the three pieces can be used to form a triangle?
14. Choose independently two numbers B and C at random from the interval [–1, 1] with uniform distribution, and consider the quadratic equation

    $$x^2 + Bx + C = 0.$$

    Find the probability that the roots of this equation

    (a) are both real.

    (b) are both positive.

Hints: (a) requires $0 \le B^2 - 4C$, (b) requires $0 \le B^2 - 4C$, $B \le 0$, $0 \le C$.

15. At the Tunbridge World's Fair, a coin toss game works as follows. Quarters are tossed onto a checkerboard. The management keeps all the quarters, but for each quarter landing entirely within one square of the checkerboard the management pays a dollar. Assume that the edge of each square is twice the diameter of a quarter, and that the outcomes are described by coordinates chosen at random. Is this a fair game?
16. Three points are chosen at random on a circle of unit circumference. What is the probability that the triangle defined by these points as vertices has three acute angles? *Hint*: One of the angles is obtuse if and only if all three points lie in the same semi-circle. Take the circumference as the interval [0, 1]. Take one point at 0 and the others at *B* and C.

17. Write a programme to choose a random number X in the interval [2, 10] 1000 times and record what fraction of the outcomes satisfy $X > 5$, what fraction satisfy $5 < X < 7$, and what fraction satisfy $x^2 - 12x + 35 > 0$. How do these results compare with Exercise 1?
18. Write a programme to choose a point (X, Y) at random in a square of side 20 inches, doing this 10,000 times, and recording what fraction of the outcomes fall within 19 inches of the centre; of these, what fraction fall between 8 and 10 inches of the centre; and, of these, what fraction fall within the first quadrant of the square. How do these results compare with those of Exercise 4?
19. Write a programme to simulate the problem describe in Exercise 7. How do the simulation results compare with the results of Exercise 7?
20. Write a programme to simulate the problem described in Exercise 12.
21. Write a programme to simulate the problem described in Exercise 16.
22. Write a programme to carry out the following experiment. A coin is tossed 100 times and the number of heads that turn up is recorded. This experiment is then repeated 1000 times. Have your programme plot a bar graph for the proportion of the 1000 experiments in which the number of heads is n, for each n in the interval [35, 65]. Does the bar graph look as though it can be fit with a normal curve?
23. Write a programme that picks a random number between 0 and 1 and computes the negative of its logarithm. Repeat this process a large number of times and plot a bar graph to give the number of times that the outcome falls in each interval of length 0 1 in [0, 10]. On this bar graph plot a graph of the density $f(x) = e^{-x}$. How well does this density fit your graph?

Probability Distributions

Simulation of Discrete Probabilities

Probability

In this chapter, we shall first consider chance experiments with a finite number of possible outcomes $w_1, w_2, \ldots, w_n$. For example, we roll a die and the possible outcomes are 1, 2, 3, 4, 5, 6 corresponding to the side that turns up. We toss a coin with possible outcomes H (heads) and T (tails).

It is frequently useful to be able to refer to an outcome of an experiment. For example, we might want to write the mathematical expression which gives the sum of four rolls of a die. To do this, we could let X_i, $i = 1, 2, 3, 4$, represent the values of the outcomes of the four rolls, and then we could write the expression

$$X_1 + X_2 + X_3 + X_4$$

for the sum of the four rolls. The X_i's are called *random variables.* A random variable is simply an expression whose value is the outcome of a particular experiment. Just as in the case of other types of variables in mathematics, random variables can take on different values.

Let X be the random variable which represents the roll of one die. We shall assign probabilities to the possible outcomes of this experiment. We do this by assigning to each outcome w_j a non-negative number $m(w_j)$ in such a way that

$$m(w_1) + m(w_2) + \dots + m(w_6) = 1.$$

The function $m(w_j)$ is called the *distribution function* of the random variable X . For the case of the roll of the die we would assign equal probabilities or probabilities 1/6 to each of the outcomes. With this assignment of probabilities, one could write

$$P(X \le 4) = \frac{2}{3}$$

to mean that the probability is 2 = 3 that a roll of a die will have a value which does not exceed 4.

Let Y be the random variable which represents the toss of a coin. In this case, there are two possible outcomes, which we can label as H and T. Unless we have reason to suspect that the coin comes up one way more often than the other way, it is natural to assign the probability of 1/2 to each of the two outcomes.

In both of the above experiments, each outcome is assigned an equal probability. This would certainly not be the case in general. For example, if a drug is found to be effective 30 per cent of the time it is used, we might assign a probability .3 that the drug is effective the next time it is used and .7 that it is not effective. This last example illustrates the intuitive frequency concept of probability. That is, if we have a probability p that an experiment will result in outcome A, then if we repeat this experiment a large number of times we should expect that the fraction of times that A will occur is about p. To check intuitive ideas like this, we shall find it helpful to look at some of these problems experimentally. We could, for example, toss a coin a large number of times and see if the fraction of times heads turns up is about 1/2. We could also simulate this experiment on a computer.

Simulation

We want to be able to perform an experiment that corresponds to a given set of probabilities; for example, $m(w_1) = 1/2$, $m(w_2) =$

1/3, and $m(w_3) = 1/6$. In this case, one could mark three faces of a six-sided die with an w_1, two faces with an w_2, and one face with an w_3.

In the general case we assume that $m(w_1)$, $m(w_2)$, . . ., $m(w_n)$ are all rational numbers, with least common denominator n. If $n > 2$, we can imagine a long cylindrical die with a cross-section that is a regular n-gon. If $m(w_j) = n_j/n$, then we can label n_j of the long faces of the cylinder with an w_j, and if one of the end faces comes up, we can just roll the die again. If $n = 2$, a coin could be used to perform the experiment.

We will be particularly interested in repeating a chance experiment a large number of times. Although the cylindrical die would be a convenient way to carry out a few repetitions, it would be difficult to carry out a large number of experiments. Since the modern computer can do a large number of operations in a very short time, it is natural to turn to the computer for this task.

Random Numbers

We must first find a computer analogue of rolling a die. This is done on the computer by means of a random number generator. Depending upon the particular software package, the computer can be asked for a real number between 0 and 1, or an integer in a given set of consecutive integers. In the first case, the real numbers are chosen in such a way that the probability that the number lies in any particular subinterval of this unit interval is equal to the length of the subinterval. In the second case, each integer has the same probability of being chosen.

Table: Sample output of the programme Random Numbers.

.203309	.762057	.151121	.623868
.932052	.415178	.716719	.967412
.069664	.670982	.352320	.049723
.750216	.784810	.089734	.966730
.946708	.380365	.027381	.900794

Let X be a random variable with distribution function $m(w)$, where w is in the set $\{w_1, w_2, w_3\}$, and $m(w_1) = 1/2$, $m(w_2) = 1/3$, and $m(w_3) = 1/6$. If our computer package can return a random integer in the set {1, 2, ..., 6}, then we simply ask it to do so, and make 1, 2, and 3 correspond to w_1, 4 and 5 correspond to w_2, and 6 correspond to w_3. If our computer package returns a random real number r in the interval (0, 1), then the expression

$$\lfloor 6r \rfloor + 1$$

will be a random integer between 1 and 6. (The notation $\lfloor x \rfloor$ means the greatest integer not exceeding x, and is read "floor of x.")

The method by which random real numbers are generated on a computer is described in the historical discussion. The following example gives sample output of the programme *Random Numbers.*

Example: (Random Number Generation) The programme *Random Numbers* generates n random real numbers in the interval [0, 1], where n is chosen by the user. When we ran the programme with $n = 20$, we obtained the data shown in the previous table.

Example: (Coin Tossing) As we have noted, our intuition suggests that the probability of obtaining a head on a single toss of a coin is 1/2. To have the computer toss a coin, we can ask it to pick a random real number in the interval [0, 1] and test to see if this number is less than 1/2. If so, we shall call the outcome heads; if not we call it tails. Another way to proceed would be to ask the computer to pick a random integer from the set {0, 1}. The programme *Coin Tosses* carries out the experiment of tossing a coin n times. Running this programme, with $n = 20$, resulted in:

THTTTHTTTTHTTTTTHHTT.

Note that in 20 tosses, we obtained 5 heads and 15 tails. Let us toss a coin n times, where n is much larger than 20, and see if we obtain a proportion of heads closer to our intuitive guess of 1/2. The programme *Coin Tosses* keeps track of the number of heads. When we ran this programme with $n = 1000$, we obtained 494 heads. When we ran it with $n = 10000$, we obtained 5039 heads.

We notice that when we tossed the coin 10,000 times, the proportion of heads was close to the "true value" .5 for obtaining a head when a coin is tossed. A mathematical model for this experiment is called *Bernoulli Trials.* The *Law of Large Numbers,* which we shall study later, will show that in the Bernoulli Trials model, the proportion of heads should be near .5, consistent with our intuitive idea of the frequency interpretation of probability.

Of course, our programme could be easily modified to simulate coins for which the probability of a head is p, where p is a real number between 0 and 1.

In the case of coin tossing, we already knew the probability of the event occurring on each experiment. The real power of simulation comes from the ability to estimate probabilities when they are not known ahead of time. This method has been used in the recent discoveries of strategies that make the casino game of blackjack favourable to the player. We illustrate this idea in a simple situation in which we can compute the true probability and see how effective the simulation is.

Example: (Dice Rolling) We consider a dice game that played an important role in the historical development of probability. The famous letters between Pascal and Fermat, which many believe started a serious study of probability, were instigated by a request for help from a French nobleman and gambler, Chevalier de Mere. It is said that de Mere had been betting that, in four rolls of a die, at least one six would turn up. He was winning consistently and, to get more people to play, he changed the game to bet that, in 24 rolls of two dice, a pair of sixes would turn up. It is claimed that de Mere lost with 24 and felt that 25 rolls were necessary to make the game favourable. It was *un grand scandale* that mathematics was wrong.

We shall try to see if de Mere is correct by simulating his various bets. The programme *De Mere1* simulates a large number of experiments, seeing, in each one, if a six turns up in four rolls of a die. When we ran this programme for 1,000 plays, a six came up in the first four rolls 48.6 per cent of the time. When we ran it for 10,000 plays this happened 51.98 per cent of the time.

We note that the result of the second run suggests that de Mere was correct in believing that his bet with one die was favourable; however, if we had based our conclusion on the first run, we would have decided that he was wrong. Accurate results by simulation require a large number of experiments.

The programme *De Mere2* simulates de Mere's second bet that a pair of sixes will occur in n rolls of a pair of dice. The previous simulation shows that it is important to know how many trials we should simulate in order to expect a certain degree of accuracy in our approximation. We shall see later that in these types of experiments, a rough rule of thumb is that, at least 95 per cent of the time, the error does not exceed the reciprocal of the square root of the number of trials. Fortunately, for this dice game, it will be easy to compute the exact probabilities. The first bet the probability that de Mere wins is $1 - (5/6)^4 = .518$.

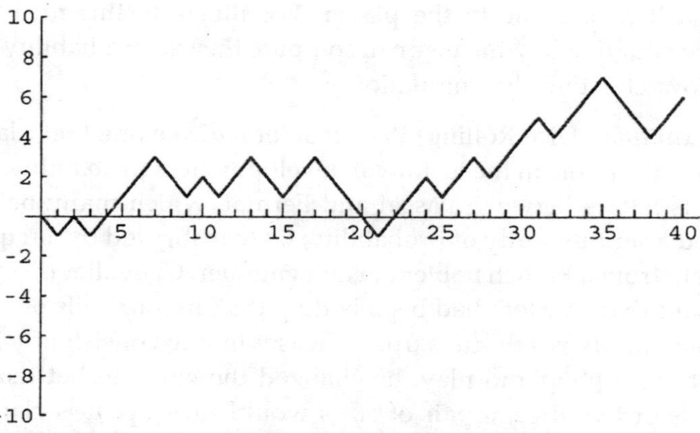

Figure: Peter's winnings in 40 plays of heads or tails.

One can understand this calculation as follows: The probability that no. 6 turns up on the first toss is (5/6). The probability that no. 6 turns up on either of the first two tosses is $(5/6)^2$. Reasoning in the same way, the probability that no 6 turns up on any of the first four tosses is $(5/6)^4$. Thus, the probability of at least one 6 in the first four tosses is $1 - (5/6)^4$. Similarly, for the second bet, with 24 rolls, the probability that de Mere wins is $1 - (35/36)^{24} = .491$, and for 25 rolls it is $1 - (35/36)^{25} = .506$.

Using the rule of thumb mentioned above, it would require 27,000 rolls to have a reasonable chance to determine these probabilities with sufficient accuracy to assert that they lie on opposite sides of .5. It is interesting to ponder whether a gambler can detect such probabilities with the required accuracy from gambling experience. Some writers on the history of probability suggest that de Mere was, in fact, just interested in these problems as intriguing probability problems.

Example: (Heads or Tails) For our next example, we consider a problem where the exact answer is difficult to obtain but for which simulation easily gives the qualitative results. Peter and Paul play a game called *heads or tails.* In this game, a fair coin is tossed a sequence of times – we choose 40. Each time a head comes up Peter wins 1 penny from Paul, and each time a tail comes up Peter loses 1 penny to Paul. For example, if the results of the 40 tosses are

THTHHHHTTHTHHTTHHTTTTHHHTHHTHHHTHHHTTTHH.

Peter's winnings may be graphed as in the previous figure.

Peter has won 6 pennies in this particular game. It is natural to ask for the probability that he will win j pennies; here j could be any even number from –40 to 40. It is reasonable to guess that the value of j with the highest probability is $j = 0$, since this occurs when the number of heads equals the number of tails.

Similarly, we would guess that the values of j with the lowest probabilities are $j = \pm 40$.

A second interesting question about this game is the following: How many times in the 40 tosses will Peter be in the lead? Looking at the graph of his winnings, we see that Peter is in the lead when his winnings are positive, but we have to make some convention when his winnings are 0 if we want all tosses to contribute to the number of times in the lead. We adopt the convention that, when Peter's winnings are 0, he is in the lead if he was ahead at the previous toss and not if he was behind at the previous toss. With this convention, Peter is in the lead 34 times in our example. Again, our intuition might suggest that the most likely number of times to be in the lead is 1/2 of 40, or 20, and the least likely numbers are the extreme cases of 40 or 0.

It is easy to settle this by simulating the game a large number of times and keeping track of the number of times that Peter's final winnings are j, and the number of times that Peter ends up being in the lead by k. The proportions overall games then give estimates for the corresponding probabilities. The programme *HTSimulation* carries out this simulation. Note that when there are an even number of tosses in the game, it is possible to be in the lead only an even number of times. We have simulated this game 10,000 times. The results are shown in the following figures. These graphs, which we call spike graphs, were generated using the programme Spikegraph. The vertical line, or spike, at position x on the horizontal axis, has a height equal to the proportion of outcomes which equal x.

Our intuition about Peter's final winnings was quite correct, but our intuition about the number of times Peter was in the lead was completely wrong. The simulation suggests that the least likely number of times in the lead is 20 and the most likely is 0 or 40. This is indeed correct, and the explanation for it is suggested by playing the game of heads or tails with a large number of tosses and looking at a graph of Peter's winnings. In the following figure, we show the results of a simulation of the game, for 1,000 tosses and in the following figure for 10,000 tosses.

In the second example Peter was ahead most of the time. It is a remarkable fact, however, that, if play is continued long enough, Peter's winnings will continue to come back to 0, but there will be very long times between the times that this happens.

In all of our examples so far, we have simulated equiprobable outcomes. We illustrate next an example where the outcomes are not equiprobable.

Example: (Horse Races) Four horses (Acorn, Balky, Chestnut, and Dolby) have raced many times. It is estimated that Acorn wins 30 per cent of the time, Balky 40 per cent of the time, Chestnut 20 per cent of the time, and Dolby 10 per cent of the time.

We can have our computer carry out one race as follows: Choose a random number x. If $x < .3$ then we say that Acorn won. If $.3 \leq x < .7$ then Balky wins. If $.7 \leq x < .9$ then Chestnut wins. Finally, if $.9 \leq x$ then Dolby wins.

The programme *Horse Race* uses this method to simulate the outcomes of n races. Running this programme for $n = 10$ we found that Acorn won 30 per cent of the time, Balky 40 per cent of the time, Chestnut 20 per cent of the time, and Dolby 10 per cent of the time. A larger number of races would be necessary to have better agreement with the past experience. Therefore, we ran the programme to simulate 1,000 races with our four horses. Although very tired after all these races, they performed in a manner quite consistent with our estimates of their abilities. Acorn won 29.8 per cent of the time, Balky 39.4 per cent, Chestnut 19.5 per cent, and Dolby 11.3 per cent of the time.

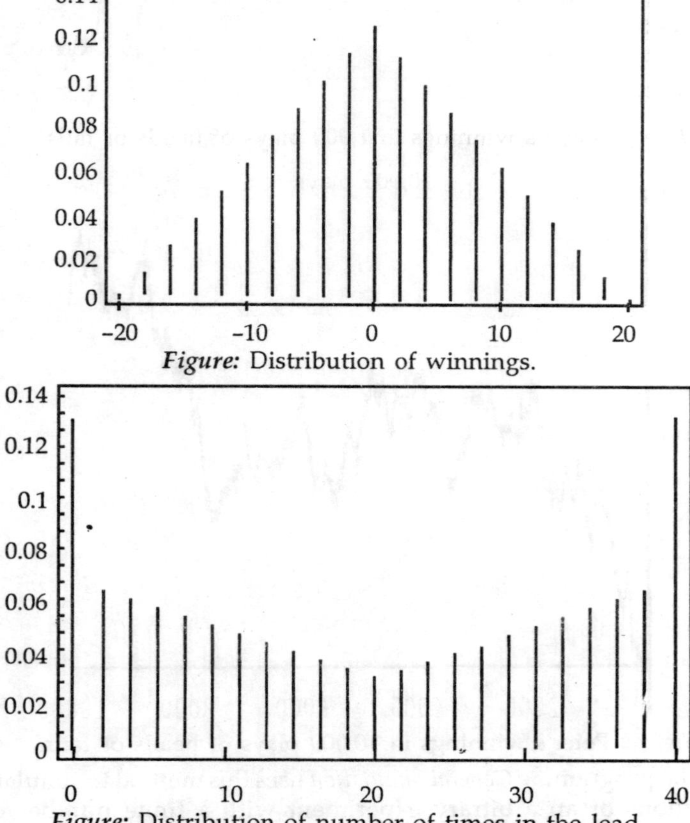

Figure: Distribution of winnings.

Figure: Distribution of number of times in the lead.

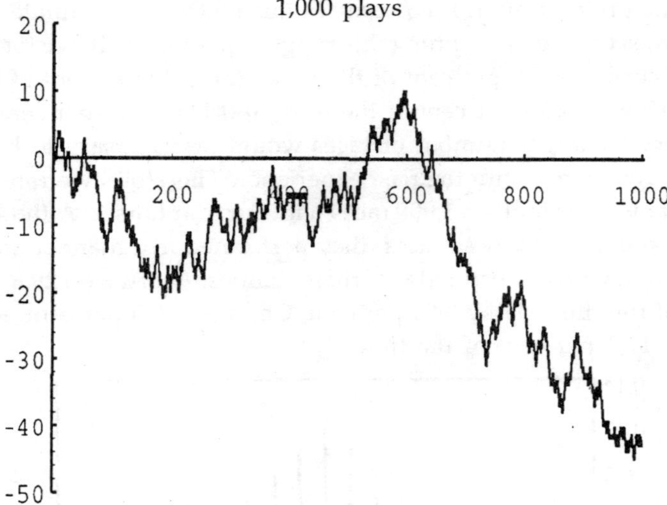

Figure: Peter's winnings in 1,000 plays of heads or tails.

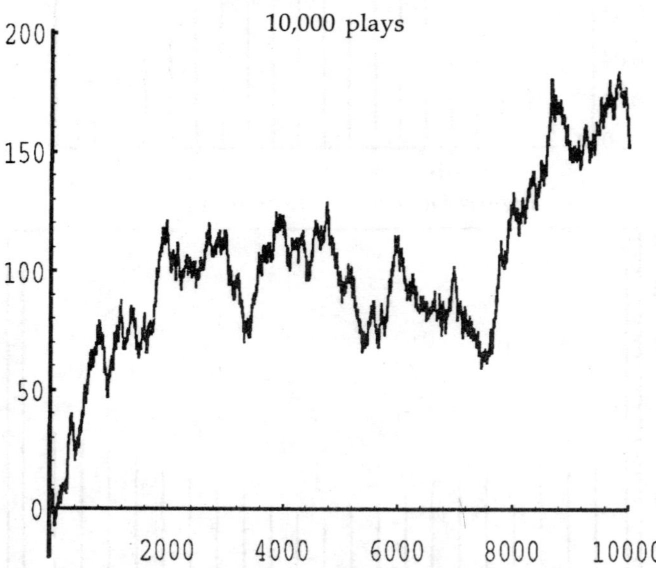

Figure: Peter's winnings in 10,000 plays of heads or tails.

The programme *General Simulation* uses this method to simulate repetitions of an arbitrary experiment with a finite number of outcomes occurring with known probabilities.

Historical Remarks

Anyone who plays the same chance game over and over is really carrying out a simulation, and in this sense the process of simulation has been going on for centuries. As we have remarked, many of the early problems of probability might well have been suggested by gamblers' experiences.

It is natural for anyone trying to understand probability theory to try simple experiments by tossing coins, rolling dice, and so forth. The naturalist Buffon tossed a coin 4040 times, resulting in 2048 heads and 1992 tails. He also estimated the number π by throwing needles on a ruled surface and recording how many times the needles crossed a line. The English biologist W. F. R. Weldon recorded 26,306 throws of 12 dice, and the Swiss scientist Rudolf Wolf recorded 100,000 throws of a single die without a computer. Such experiments are very time-consuming and may not accurately represent the chance phenomena being studied. For example, for the dice experiments of Weldon and Wolf, further analysis of the recorded data showed a suspected bias in the dice. The statistician Karl Pearson analysed a large number of outcomes at certain roulette tables and suggested that the wheels were biased. He wrote in 1894:

> "Clearly, since the Casino does not serve the valuable end of huge laboratory for the preparation of probability statistics, it has no scientific *raison d'etre*. Men of science cannot have their most refined theories disregarded in this shameless manner! The French Government must be urged by the hierarchy of science to close the gaming-saloons; it would be, of course, a graceful act to hand over the remaining resources of the Casino to the Academie des Sciences for the endowment of a laboratory of orthodox probability; in particular, of the new branch of that study, the application of the theory of chance to the biological problems of evolution, which is likely to occupy so much of men's thoughts in the near future."

However, these early experiments were suggestive and led to important discoveries in probability and statistics. They led Pearson to the *chi-squared test,* which is of great importance in testing whether observed data fit a given probability distribution.

By the early-1900s it was clear that a better way to generate random numbers was needed. In 1927, L. H. C. Tippett published a list of 41,600 digits obtained by selecting numbers haphazardly from census reports. In 1955, RAND Corporation printed a table of 1,000,000 random numbers generated from electronic noise. The advent of the high-speed computer raised the possibility of generating random numbers directly on the computer, and in the late-1940s John von Neumann suggested that this be done as follows: Suppose that you want a random sequence of four-digit numbers. Choose any four-digit number, say 6235, to start. Square this number to obtain 38,875,225. For the second number choose the middle four digits of this square (i.e., 8752). Do the same process starting with 8752 to get the third number, and so forth.

More modern methods involve the concept of modular arithmetic. If a is an integer and m is a positive integer, then by a (mod m) we mean the remainder when a is divided by m. For example, 10 (mod 4) = 2, 8 (mod 2) = 0, and so forth. To generate a random sequence X_0, X_1, X_2, ... of numbers choose a starting number X_0 and then obtain the numbers X_{n+1} from X_n by the formula

$$X_{n+1} = (aX_n + c) \pmod{m},$$

where a, c, and m are carefully chosen constants. The sequence X_0, X_1, X_2, ... is then a sequence of integers between 0 and $m - 1$. To obtain a sequence of real numbers in (0, 1), we divide each X_j by m. The resulting sequence consists of rational numbers of the form $j = m$, where $0 \leq j \leq m - 1$. Since m is usually a very large integer, we think of the numbers in the sequence as being random real numbers in [0, 1).

For both von Neumann's squaring method and the modular arithmetic technique the sequence of numbers is actually completely determined by the first number. Thus, there is nothing really random about these sequences. However, they produce numbers that behave very much as theory would predict for random experiments. To obtain different sequences for different experiments the initial number X_0 is chosen by some other procedure that might involve, for example, the time of day.

During the Second World War, physicists at the Los Alamos Scientific Laboratory needed to know, for purposes of shielding, how far neutrons travel through various materials. This question was beyond the reach of theoretical calculations. Daniel McCracken, writing in the Scientific American, states:

> "The physicists had most of the necessary data: they knew the average distance a neutron of a given speed would travel in a given substance before it collided with an atomic nucleus, what the probabilities were that the neutron would bounce off instead of being absorbed by the nucleus, how much energy the neutron was likely to lose after a given collision and so on."

John von Neumann and Stanislas Ulam suggested that the problem be solved by modelling the experiment by chance devices on a computer. Their work being secret, it was necessary to give it a code name. Von Neumann chose the name "Monte Carlo." Since that time, this method of simulation has been called the *Monte Carlo Method.*

William Feller indicated the possibilities of using computer simulations to illustrate basic concepts in probability in his book, An Introduction to Probability Theory and its Applications. In discussing the problem about the number of times in the lead in the game of "heads or tails" Feller writes:

> "The results concerning fluctuations in coin tossing show that widely held beliefs about the law of large numbers are fallacious. These results are so amazing and so at variance with common intuition that even sophisticated colleagues doubted that coins actually misbehave as theory predicts. The record of a simulated experiment is therefore included."

Feller provides a plot showing the result of 10,000 plays of heads or tails similar to that in the previous figure.

The martingale betting system has a long and interesting history. Russell Barnhart pointed out to the authors that its use can be traced back at least to 1754, when Casanova, writing in his memoirs, History of My Life, writes:

> "She (Casanova's mistress) made me promise to go to the casino (the Ridotto in Venice) for money to play in partnership with her. I went there and took all the gold I found, and determinedly doubling my stakes according to the system known as the *martingale*, I won three or four times a day during the rest of the Carnival. I never lost the sixth card. If I had lost it, I should have been out of funds, which amounted to two thousand zecchini."

Even if there were no zeros on the roulette wheel so the game was perfectly fair, the martingale system, or any other system for that matter, cannot make the game into a favourable game. The idea that a fair game remains fair and unfair games remain unfair under gambling systems has been exploited by mathematicians to obtain important results in the study of probability.

The word 'martingale' itself also has an interesting history. The origin of the word is obscure. The Oxford English Dictionary gives examples of its use in the early-1600s and says that its probable origin is the reference in Rabelais's Book One:

> "Everything was done as planned, the only thing being that Gargantua doubted if they would be able to find, right away, breeches suitable to the old fellow's legs; he was doubtful, also, as to what cut would be most becoming to the orator – the martingale, which has a draw-bridge effect in the seat, to permit doing one's business more easily; the sailor-style, which affords more comfort for the kidneys; the Swiss, which is warmer on the belly; or the codfish-tail, which is cooler on the loins."

In modern uses martingale has several different meanings, all related to holding down, in addition to the gambling use. For example, it is a strap on a horse's harness used to hold down the horse's head, and also part of a sailing rig used to hold down the bowsprit.

The Labouchere system is named after Henry du Pre Labouchere (1831-1912), an English journalist and member of Parliament. Labouchere attributed the system to Condorcet. Condorcet (1743-1794) was a political leader during the time of the French revolution who was interested in applying probability theory to economics and politics. For example, he calculated the probability that a jury using majority vote will give a correct decision if each juror has the same probability of deciding correctly. His writings provided a wealth of ideas on how probability might be applied to human affairs.

Exercises

1. Modify the programme *Coin Tosses* to toss a coin n times and print out after every 100 tosses the proportion of heads minus 1/2. Do these numbers appear to approach 0 as n increases? Modify the programme again to print out, every 100 times, both of the following quantities: the proportion of heads minus 1/2, and the number of heads minus half the number of tosses. Do these numbers appear to approach 0 as n increases?
2. Modify the programme *Coin Tosses* so that it tosses a coin n times and records whether or not the proportion of heads is within .1 of .5 (i.e., between .4 and .6). Have your programme repeat this experiment 100 times. About how large must n be so that approximately 95 out of 100 times the proportion of heads is between .4 and .6?
3. In the early-1600s, Galileo was asked to explain the fact that, although the number of triples of integers from 1 to 6 with sum 9 is the same as the number of such triples with sum 10, when three dice are rolled, a 9 seemed to come up less often than a 10—supposedly in the experience of gamblers.
   (a) Write a programme to simulate the roll of three dice a large number of times and keep track of the proportion of times that the sum is 9 and the proportion of times it is 10.
   (b) Can you conclude from your simulations that the gamblers were correct?

4. In raquetball, a player continues to serve as long as she is winning; a point is scored only when a player is serving and wins the volley. The first player to win 21 points wins the game. Assume that you serve first and have a probability .6 of winning a volley when you serve and probability .5 when your opponent serves. Estimate, by simulation, the probability that you will win a game.
5. Consider the bet that all three dice will turn up sixes at least once in n rolls of three dice. Calculate $f(n)$, the probability of at least one triple-six when three dice are rolled n times. Determine the smallest value of n necessary for a favourable bet that a triple-six will occur when three dice are rolled n times.

   (*DeMoivre* would say it should be about 216 log 2 = 149:7 and so would answer 150. Do you agree with him?)
6. In Las Vegas, a roulette wheel has 38 slots numbered 0, 00, 1, 2, . . ., 36. The 0 and 00 slots are green and half of the remaining 36 slots are red and half are black. A croupier spins the wheel and throws in an ivory ball. If you bet 1 dollar on red, you win 1 dollar if the ball stops in a red slot and otherwise you lose 1 dollar. Write a programme to find the total winnings for a player who makes 1,000 bets on red.
7. Another form of bet for roulette is to bet that a specific number (say 17) will turn up. If the ball stops on your number, you get your dollar back plus 35 dollars. If not, you lose your dollar. Write a programme that will plot your winnings when you make 500 plays of roulette at Las Vegas, first when you bet each time on red, and then for a second visit to Las Vegas when you make 500 plays betting each time on the number 17. What differences do you see in the graphs of your winnings on these two occasions?
8. An astute student noticed that, in our simulation of the game of heads or tails, the proportion of times the player is always in the lead is very close to the proportion of times that the player's total winnings end up 0.

Work out these probabilities by enumeration of all cases for two tosses and for four tosses, and see if you think that these probabilities are, in fact, the same.

9. The *Labouchere system* for roulette is played as follows. Write down a list of numbers, usually 1, 2, 3, 4. Bet the sum of the first and last, 1 + 4 = 5, on red. If you win, delete the first and last numbers from your list. If you lose, add the amount that you last bet to the end of your list. Then use the new list and bet the sum of the first and last numbers (if there is only one number, bet that amount). Continue until your list becomes empty. Show that, if this happens, you win the sum, 1 + 2 + 3 + 4 = 10, of your original list. Simulate this system and see if you do always stop and, hence, always win. If so, why is this not a foolproof gambling system?
10. Another well-known gambling system is the *martingale doubling system*. Suppose that you are betting on red to turn up in roulette. Every time you win, bet 1 dollar next time. Every time you lose, double your previous bet. Continue to play until you have won at least 5 dollars or you have lost more than 100 dollars. Write a programme to simulate this system and play it a number of times and see how you do. In his book, The Newcomes, W. M. Thackeray remarks, "You have not played as yet? Do not do so; above all avoid a martingale if you do." Was this good advice?
11. Modify the programme *HT Simulation* so that it keeps track of the maximum of Peter's winnings in each game of 40 tosses. Have your programme print out the proportion of times that your total winnings take on values 0, 2, 4, ..., 40. Calculate the corresponding exact probabilities for games of two tosses and four tosses.
12. In an upcoming national election for the President of the United States, a pollster plans to predict the winner of the popular vote by taking a random sample of 1,000 voters and declaring that the winner will be the one obtaining the most votes in his sample. Suppose that 48 per cent of the voters plan to vote for the Republican candidate and 52 per cent

plan to vote for the Democratic candidate. To get some idea of how reasonable the pollster's plan is, write a programme to make this prediction by simulation. Repeat the simulation 100 times and see how many times the pollster's prediction would come true. Repeat your experiment, assuming now that 49 per cent of the population plan to vote for the Republican candidate; first with a sample of 1,000 and then with a sample of 3,000. (The Gallup Poll uses about 3,000).

13. The psychologist Tversky and his colleagues say that about four out of five people will answer (a) to the following question:

    A certain town is served by two hospitals. In the larger hospital about 45 babies are born each day, and in the smaller hospital 15 babies are born each day. Although the overall proportion of boys is about 50 per cent, the actual proportion at either hospital may be more or less than 50 per cent on any day. At the end of a year, which hospital will have the greater number of days on which more than 60 per cent of the babies born were boys?

    (a) the large hospital.

    (b) the small hospital.

    (c) neither—the number of days will be about the same.

    Assume that the probability that a baby is a boy is .5 (actual estimates make this more like .513). Decide, by simulation, what the right answer is to the question. Can you suggest why so many people go wrong?

14. You are offered the following game. A fair coin will be tossed until the first time it comes up heads. If this occurs on the jth toss you are paid 2^j dollars.

    You are sure to win at least 2 dollars so you should be willing to pay to play this game—but how much? Few people would pay as much as 10 dollars to play this game. See if you can decide, by simulation, a reasonable amount that you would be willing to pay, per game, if you will be allowed to make a large number of plays of the game. Does the amount that you would be willing to pay per game depend upon the number of plays that you will be allowed?

15. Tversky and his colleagues studied the records of 48 of the Philadelphia 76ers basketball games in the 1980-81 season to see if a player had times when he was hot and every shot went in, and other times when he was cold and barely able to hit the backboard. The players estimated that they were about 25 per cent more likely to make a shot after a hit than after a miss.

    In fact, the opposite was true—the 76ers were 6 per cent more likely to score after a miss than after a hit. Tversky reports that the number of hot and cold streaks was about what one would expect by purely random effects. Assuming that a player has a fifty-fifty chance of making a shot and makes 20 shots a game, estimate by simulation the proportion of the games in which the player will have a streak of 5 or more hits.

16. Estimate, by simulation, the average number of children there would be in a family if all people had children until they had a boy. Do the same if all people had children until they had at least one boy and at least one girl. How many more children would you expect to find under the second scheme than under the first in 100,000 families? (Assume that boys and girls are equally likely.)

17. Mathematicians have been known to get some of the best ideas while sitting in a cafe, riding on a bus, or strolling in the park. In the early-1900s, the famous mathematician George Polya lived in a hotel near the woods in Zurich. He liked to walk in the woods and think about mathematics. Polya describes the following incident:

"At the hotel there lived also some students with whom I usually took my meals and had friendly relations. On a certain day one of them expected the visit of his fiancee, what (sic) I knew, but I did not foresee that he and his fiancee would also set out for a stroll in the woods, and then suddenly I met them there. And then I met them the same morning repeatedly, I don't remember how many times, but certainly much too often and I felt embarrassed: It looked as if I was snooping around which was, I assure you, not the case."

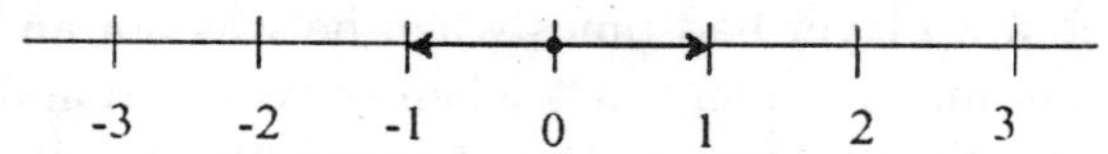

a. Random walk in one dimension.

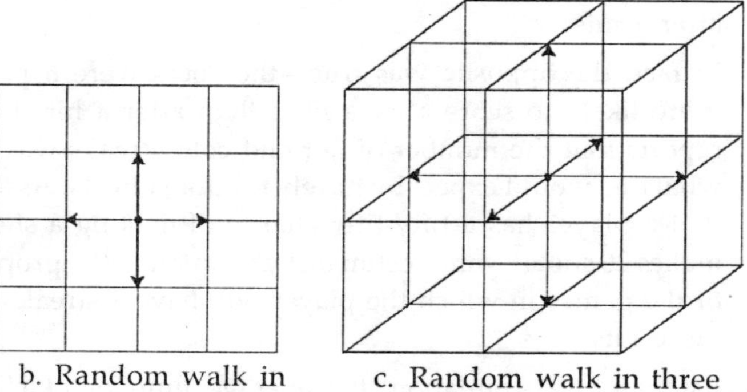

b. Random walk in two dimensions.

c. Random walk in three dimensions.

Figure: Random walk.

This set him to thinking about whether random walkers were destined to meet.

Polya considered random walkers in one, two, and three dimensions. In one dimension, he envisioned the walker on a very long street. At each intersection the walker flips a fair coin to decide which direction to walk next. In two dimensions, the walker is walking on a grid of streets, and at each intersection he chooses one of the four possible directions with equal probability. In three dimensions (we might better speak of a random climber), the walker moves on a three-dimensional grid, and at each intersection there are now six different directions that the walker may choose, each with equal probability.

(a) Write a programme to simulate a random walk in one dimension starting at 0. Have your programme print out the lengths of the times between returns to the starting point (returns to 0). See if you can guess from this simulation the answer to the following question: Will the walker always return to his starting point eventually or might he drift away forever?

(b) The paths of two walkers in two dimensions who meet after n steps can be considered to be a single path that starts at (0, 0) and returns to (0, 0) after $2n$ steps. This means that the probability that two random walkers in two dimensions meet is the same as the probability that a single walker in two dimensions ever returns to the starting point. Thus, the question of whether two walkers are sure to meet is the same as the question of whether a single walker is sure to return to the starting point.

Write a programme to simulate a random walk in two dimensions and see if you think that the walker is sure to return to (0, 0). If so, Polya would be sure to keep meeting his friends in the park. Perhaps by now you have conjectured the answer to the question: Is a random walker in one or two dimensions sure to return to the starting point? Polya answered this question for dimensions one, two, and three. He established the remarkable result that the answer is yes in one and two dimensions and no in three dimensions.

(c) Write a programme to simulate a random walk in three dimensions and see whether, from this simulation and the results of (a) and (b), you could have guessed Polya's result.

Discrete Probability Distributions

In this book we shall study many different experiments from a probabilistic point of view. What is involved in this study will become evident as the theory is developed and examples are analysed. However, the overall idea can be described and illustrated as follows: to each experiment that we consider there will be associated a random variable, which represents the outcome of any particular experiment. The set of possible outcomes is called the *sample space*. In the first part of this section, we will consider the case where the experiment has only finitely many possible outcomes, i.e., the sample space is finite. We will then generalise to the case that the sample space is either finite or countably infinite. This leads us to the following definition.

Random Variables and Sample Spaces

Definition: Suppose we have an experiment whose outcome depends on chance. We represent the outcome of the experiment by a capital Roman letter, such as X, called a *random variable*. The sample space of the experiment is the set of all possible outcomes. If the sample space is either finite or countably infinite, the random variable is said to be discrete.

We generally denote a sample space by the capital Greek letter Ω. As stated above, in the correspondence between an experiment and the mathematical theory by which it is studied, the sample space Ω corresponds to the set of possible outcomes of the experiment.

We now make two additional definitions. These are subsidiary to the definition of sample space and serve to make precise some of the common terminology used in conjunction with sample spaces. First of all, we define the elements of a sample space to be outcomes. Second, each subset of a sample space is defined to be an event. Normally, we shall denote outcomes by lower case letters and events by capital letters.

Example: A die is rolled once. We let X denote the outcome of this experiment. Then the sample space for this experiment is the 6-element set

$$\Omega = \{1, 2, 3, 4, 5, 6\},$$

where each outcome i, for $i = 1, \ldots, 6$, corresponds to the number of dots on the face which turns up. The event

$$E = \{2, 4, 6\}$$

corresponds to the statement that the result of the roll is an even number. The event E can also be described by saying that X is even. Unless there is reason to believe the die is loaded, the natural assumption is that every outcome is equally likely. Adopting this convention means that we assign a probability of 1/6 to each of the six outcomes, i.e., $m(i) = 1/6$, for $1 \leq i \leq 6$.

Distribution Functions

We next describe the assignment of probabilities. The definitions are motivated by the example above, in which we assigned to each outcome of the sample space a non-negative number such that the sum of the numbers assigned is equal to 1.

Definition: Let X be a random variable which denotes the value of the outcome of a certain experiment, and assume that this experiment has only finitely many possible outcomes. Let Ω be the sample space of the experiment (i.e., the set of all possible values of X, or equivalently, the set of all possible outcomes of the experiment.) A distribution function for X is a real-valued function m whose domain is Ω and which satisfies:

1. $m(w) \geq 0$; for all $w \in \Omega$, and
2. $\sum_{w \in \Omega} m(w) = 1$.

For any subset E of Ω, we define the probability of E to be the number $P(E)$ given by

$$P(E) = \sum_{w \in W} m(w).$$

Example: Consider an experiment in which a coin is tossed twice. Let X be the random variable which corresponds to this experiment. We note that there are several ways to record the outcomes of this experiment. We could, for example, record the two tosses, in the order in which they occurred. In this case, we have Ω ={HH, HT, TH, TT}. We could also record the outcomes by simply noting the number of heads that appeared. In this case, we have Ω ={0, 1, 2}. Finally, we could record the two outcomes, without regard to the order in which they occurred. In this case, we have Ω ={HH, HT, TT}.

We will use, for the moment, the first of the sample spaces given above. We will assume that all four outcomes are equally likely, and define the distribution function $m(w)$ by

$$m(\text{HH}) = m(\text{HT}) = m(\text{TH}) = m(\text{TT}) = \frac{1}{4}.$$

Let E ={HH, HT, TH} be the event that at least one head comes up. Then, the probability of E can be calculated as follows:

$$P(E) = m(\text{HH}) + m(\text{HT}) + m(\text{TH})$$

$$= \frac{1}{4}+\frac{1}{4}+\frac{1}{4}=\frac{3}{4}$$

Similarly, if F={HH,HT} is the event that heads comes up on the first toss, then we have

$$P(F) = m(\text{HH}) + m(\text{HT})$$

$$= \frac{1}{4}+\frac{1}{4}=\frac{1}{2}$$

Example: The sample space for the experiment in which the die is rolled is the 6-element set $\Omega = \{1, 2, 3, 4, 5, 6\}$. We assumed that the die was fair, and we chose the distribution function defined by

$$m(i) = \frac{1}{6}, \text{ for } i = 1, \ldots, 6.$$

If E is the event that the result of the roll is an even number, then $E = \{2, 4, 6\}$ and

$$P(E) = m(2) + m(4) + m(6)$$

$$= \frac{1}{6}+\frac{1}{6}+\frac{1}{6}=\frac{1}{2}.$$

Notice that it is an immediate consequence of the above definitions that, for every $\omega \in \Omega$,

$$P(\{w\}) = m(w).$$

That is, the probability of the elementary event $\{w\}$, consisting of a single outcome w, is equal to the value $m(w)$ assigned to the outcome w by the distribution function.

Example: Three people, A, B, and C, are running for the same office, and we assume that one and only one of them wins. The sample space may be taken as the 3-element set $\Omega = \{A, B, C\}$ where each element corresponds to the outcome of that candidate's winning. Suppose that A and B have the same chance of winning, but that C has only 1/2 the chance of A or B. Then we assign

$$m(A) = m(B) = 2m(C).$$

Since

$$m(A) + m(B) + m(C) = 1,$$

we see that

$$2m(C) + 2m(C) + m(C) = 1,$$

which implies that $5m(C) = 1$. Hence,

$$m(A) = \frac{2}{5},\ m(B) = \frac{2}{5},\ m(C) = \frac{1}{5}.$$

Let E be the event that either A or C wins. Then $E = \{A, C\}$, and

$$P(E) = m(A) + m(C) = \frac{2}{5} + \frac{1}{5} = \frac{3}{5}.$$

In many cases, events can be described in terms of other events through the use of the standard constructions of set theory. We will briefly review the definitions of these constructions. The reader is referred to the following figure for Venn diagrams which illustrate these constructions.

Let A and B be two sets. Then the union of A and B is the set

$$A \cup B = \{x \mid x \in A \text{ or } x \in B\}.$$

The intersection of A and B is the set

$$A \cap B = \{x \mid x \in A \text{ and } x \in B\}.$$

The difference of A and B is the set

$$A - B = \{x \mid x \in A \text{ and } x \notin B\}.$$

The set A is a subset of B, written $A \subset B$, if every element of A is also an element of B . Finally, the complement of A is the set

$$\tilde{A} = \{x \mid x \in \Omega \text{ and } x \notin A\}.$$

The reason that these constructions are important is that it is typically the case that complicated events described in English can be broken down into simpler events using these constructions.

For example, if A is the event that "it will snow tomorrow and it will rain the next day," B is the event that "it will snow tomorrow," and C is the event that "it will rain two days from now," then A is the intersection of the events B and C. Similarly, if D is the event that "it will snow tomorrow or it will rain the next day," then $D = B \cup C$. (Note that care must be taken here, because sometimes the word "or" in English means that exactly one of the two alternatives will occur.

The meaning is usually clear from context. In this book, we will always use the word "or" in the inclusive sense, i.e., A or B means that at least one of the two events A, B is true.) The event $\tilde{B}$ is the event that "it will not snow tomorrow." Finally, if E is the event that "it will snow tomorrow but it will not rain the next day," then $E = B - C$.

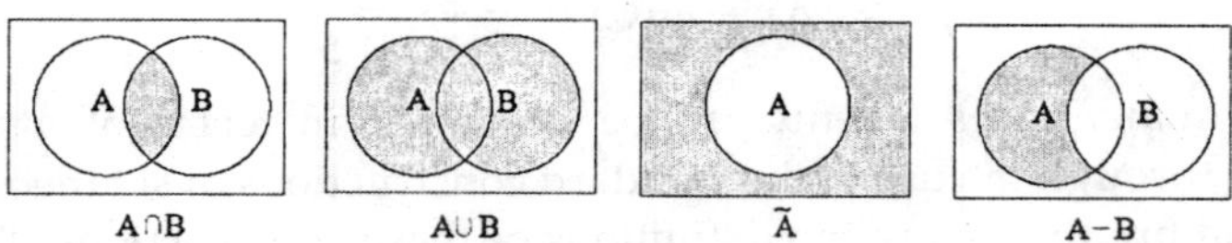

Figure: Basic set operations.

Properties

Theorem (a): The probabilities assigned to events by a distribution function on a sample space Ω satisfy the following properties:

1. $P(E) \geq 0$ for every $E \subset \Omega$.
2. $P(\Omega) = 1$.
3. If $E \subset F \subset \Omega$, then $P(E) \leq P(F)$.
4. If A and B are *disjoint* subsets of Ω, then $P(A \cup B) = P(A) + P(B)$.
5. $P(\tilde{A}) = 1 - P(A)$ for every $A \subset \Omega$.

Proof: For any event E the probability $P(E)$ is determined from the distribution m by

$$P(E) = \sum_{w \in E} m(w)$$

for every $E \subset \Omega$. Since the function m is non-negative, it follows that $P(E)$ is also non-negative. Thus, Property 1 is true.

Property 2 is proved by the equations

$$P(\Omega) = \sum_{w \in \Omega} m(w) = 1$$

Suppose that $E \subset F \subset \Omega$. Then every element w that belongs to E also belongs to F. Therefore,

$$\sum_{w \in E} m(w) \leq \sum_{w \in F} m(w)$$

since each term in the left-hand sum is in the right-hand sum, and all the terms in both sums are non-negative. This implies that

$$P(E) \leq P(F),$$

and Property 3 is proved.

Suppose next that A and B are disjoint subsets of Ω. Then every element w of $A \cup B$ lies either in A and not in B or in B and not in A. It follows that

$$\begin{aligned} P(A \cup B) &= \sum_{w \in A \cup B} m(w) = \sum_{w \in A} m(w) + \sum_{w \in B} m(w) \\ &= P(A) + P(B), \end{aligned}$$

and Property 4 is proved.

Finally, to prove Property 5, consider the disjoint union

$$\Omega = A \cup \tilde{A}$$

Since $P(\Omega) = 1$, the property of disjoint additivity (Property 4) implies that

$$1 = P(A) + P(\tilde{A}),$$

whence $P(\tilde{A}) = 1 - P(A)$.

It is important to realise that Property 4 in Theorem (a) can be extended to more than two sets. The general finite additivity property is given by the following theorem.

Theorem (b): If $A_1, \ldots, A_n$ are pairwise disjoint subsets of Ω (i.e., no two of the A_i's have an element in common), then

$$P(A_1 \cup \ldots \cup A_n) = \sum_{i=1}^{n} P(A_i).$$

Proof: Let w be any element in the union

$$A_1 \cup \ldots \cup A_n.$$

Then $m(w)$ occurs exactly once on each side of the equality in the statement of the theorem.

We shall often use the following consequence of the above theorem.

Theorem (c): Let $A_1, \ldots, A_n$ be pairwise disjoint events with $\Omega = A_1 \cup \ldots \cup A_n$, and let E be any event. Then

$$P(E) = \sum_{i=1}^{n} P(E \cap A_i).$$

Proof: The sets $E \cap A_1, \ldots, E \cap A_n$ are pairwise disjoint, and their union is the set E. The result now follows from Theorem (b).

Corollary: For any two events A and B,

$$P(A) = P(A \cap B) + P(A \cap \tilde{B}).$$

Property 4 can be generalised in another way. Suppose that A and B are subsets of Ω which are not necessarily disjoint. Then:

Theorem (d): If A and B are subsets of Ω, then

$$P(A \cup B) = P(A) + P(B) - P(A \cap B).$$

Proof: The left side of above equation is the sum of $m(w)$ for w in either A or B. We must show that the right side of above equation also adds $m(w)$ for w in A or B. If w is in exactly one of the two sets, then it is counted in only one of the three terms on the right side of above equation. If it is in both A and B, it is added twice from the calculations of $P(A)$ and $P(B)$ and subtracted once for $P(A \cap B)$. Thus, it is counted exactly once by the right side. Of course, if $A \cap B = \phi$, then above equation reduces to Property 4.

Tree Diagrams

Example: Let us illustrate the properties of probabilities of events in terms of three tosses of a coin. When we have an experiment which takes place in stages such as this, we often find it convenient to represent the outcomes by a *tree diagram* as shown in the following figure.

A path through the tree corresponds to a possible outcome of the experiment. For the case of three tosses of a coin, we have eight paths $w_1, w_2, \ldots, w_8$ and, assuming each outcome to be equally likely, we assign equal weight, 1/8, to each path. Let E be the event "at least one head turns up." Then $\tilde{E}$ is the event "no heads turn up." This event occurs for only one outcome, namely, w_8 = TTT. Thus, $\tilde{E}$ = {TTT} and we have

$$P(\tilde{E}) = P(\{\text{TTT}\}) = \text{m(TTT)} = \frac{1}{8}.$$

By Property 5 of Theorem (a),

$$P(E) = 1 - P(\tilde{E}) = 1 - \frac{1}{8} = \frac{7}{8}.$$

Note that we shall often find it is easier to compute the probability that an event does not happen rather than the probability that it does. We then use Property 5 to obtain the desired probability.

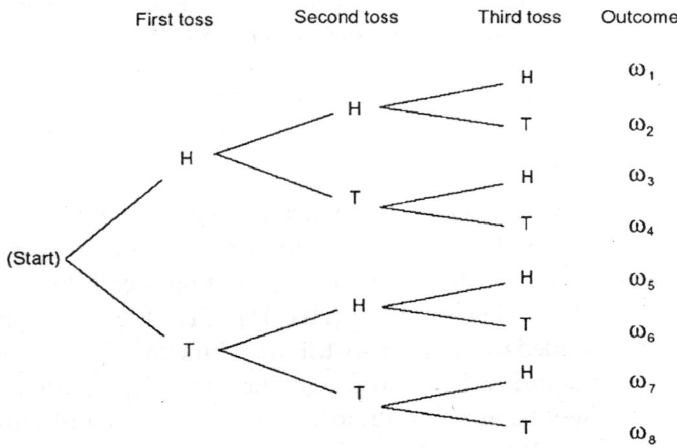

Figure: Tree diagram for three tosses of a coin.

Let A be the event "the first outcome is a head," and B the event "thesecond outcome is a tail." By looking at the paths in the above figure, we see that

$$P(A) = P(B) = \frac{1}{2}.$$

Moreover, $A \cap B = \{w_3, w_4\}$, and so $P(A \cap B) = 1/4$. Using Theorem (d), we obtain

$$P(A \cup B) = P(A) + P(B) - P(A \cap B)$$

$$= \frac{1}{2} + \frac{1}{2} - \frac{1}{4} = \frac{3}{4}.$$

Since $A \cup B$ is the 6-element set,

$A \cup B$ = {HHH, HHT, HTH, HTT, TTH, TTT},

we see that we obtain the same result by direct enumeration.

In our coin tossing examples and in the die rolling example, we have assigned an equal probability to each possible outcome of the experiment. Corresponding to this method of assigning probabilities, we have the following definitions.

Uniform Distribution

Definition: The uniform distribution on a sample space Ω containing n elements is the function m defined by

$$m(w) = \frac{1}{n},$$

for every $w \in \Omega$.

It is important to realise that when an experiment is analysed to describe its possible outcomes, there is no single correct choice of sample space. For the experiment of tossing a coin twice, we selected the 4-element set Ω = {HH, HT, TH, TT} as a sample space and assigned the uniform distribution function. These choices are certainly intuitively natural. On the other hand, for some purposes it may be more useful to consider the 3-element sample space $\bar{\Omega} = \{0, 1, 2\}$ in which 0 is the outcome "no heads turn up,"

1 is the outcome "exactly one head turns up," and 2 is the outcome "two heads turn up." The distribution function $\bar{m}$ on $\bar{\Omega}$ defined by the equations

$$\bar{m}(0)=\frac{1}{4},\quad \bar{m}(1)=\frac{1}{2},\quad \bar{m}(2)=\frac{1}{4}$$

is the one corresponding to the uniform probability density on the original sample space Ω. Notice that it is perfectly possible to choose a different distribution function. For example, we may consider the uniform distribution function on $\bar{\Omega}$, which is the function $\bar{q}$ defined by

$$\bar{q}(0) = \bar{q}(1)=\bar{q}(2)=\frac{1}{3}.$$

Although $\bar{q}$ is a perfectly good distribution function, it is not consistent with observed data on coin tossing.

Example: Consider the experiment that consists of rolling a pair of dice. We take as the sample space Ω the set of all ordered pairs (i, j) of integers with $1 \le i \le 6$ and $1 \le j \le 6$. Thus,

$$\Omega = \{(i, j): 1 \le i, j \le 6\}.$$

(There is at least one other "reasonable" choice for a sample space, namely the set of all unordered pairs of integers, each between 1 and 6.) To determine the size of Ω, we note that there are six choices for i, and for each choice of i there are six choices for j, leading to 36 different outcomes. Let us assume that the dice are not loaded. In mathematical terms, this means that we assume that each of the 36 outcomes is equally likely, or equivalently, that we adopt the uniform distribution function on Ω by setting

$$m(i, j) = \frac{1}{36},\quad 1 \le i, j \le 6.$$

What is the probability of getting a sum of 7 on the roll of two dice—or getting a sum of 11? The first event, denoted by E, is the subset

$$E = \{(1, 6), (6, 1), (2, 5), (5, 2), (3, 4), (4, 3)\}.$$

A sum of 11 is the subset F given by

$$F = \{(5, 6), (6, 5)\}.$$

Consequently,

$$P(E) = \sum_{w\in E} m(w) = 6\cdot\frac{1}{36} = \frac{1}{6},$$

$$P(F) = \sum_{w\in F} m(w) = 2\cdot\frac{1}{36} = \frac{1}{18}.$$

What is the probability of getting neither snake-eyes (double ones) nor boxcars (double sixes)? The event of getting either one of these two outcomes is the set

$$E = \{(1, 1), (6, 6)\}.$$

Hence, the probability of obtaining neither is given by

$$P(\tilde{E}) = 1 - P(E) = 1 - \frac{2}{36} = \frac{17}{18}.$$

In the above coin tossing and the dice rolling experiments, we have assigned an equal probability to each outcome. That is, in each example, we have chosen the uniform distribution function. These are the natural choices provided the coin is a fair one and the dice are not loaded. However, the decision as to which distribution function to select to describe an experiment is not a part of the basic mathematical theory of probability. The latter begins only when the sample space and the distribution function have already been defined.

Determination of Probabilities

It is important to consider ways in which probability distributions are determined in practice. One way is by symmetry. For the case of the toss of a coin, we do not see any physical difference between the two sides of a coin that should affect the chance of one side or the other turning up. Similarly, with an ordinary die there is no essential difference between any two sides of the die, and so by symmetry we assign the same probability for any possible outcome. In general, considerations of symmetry often suggest the uniform distribution function. Care must be used here. We should not always assume that, just because we do not know any reason to suggest that one outcome is more likely than another, it is appropriate to assign equal probabilities. For example, consider the experiment of guessing the sex of a newborn child. It has been observed that the proportion of newborn children who are boys is about .513. Thus, it is more appropriate to assign a distribution function which assigns probability .513 to the outcome boy and probability .487 to the outcome girl than to assign probability 1/2 to each outcome. This is an example where we use statistical observations to determine probabilities. Note that these probabilities may change with new studies and may vary from country to country. Genetic engineering might even allow an individual to influence this probability for a particular case.

Odds

Statistical estimates for probabilities are fine if the experiment under consideration can be repeated a number of times under similar circumstances. However, assume that, at the beginning of a football season, you want to assign a probability to the event that Dartmouth will beat Harvard. You really do not have data that relates to this year's football team. However, you can determine your own personal probability by seeing what kind of a bet you would be willing to make. For example, suppose that you are willing to make a 1 dollar bet giving 2 to 1 odds that Dartmouth will win. Then you are willing to pay 2 dollars if Dartmouth loses in return for receiving 1 dollar if Dartmouth wins. This means that you think the appropriate probability for Dartmouth winning is 2/3.

Let us look more carefully at the relation between odds and probabilities. Suppose that we make a bet at r to 1 odds that an event E occurs. This means that we think that it is r times as likely that E will occur as that E will not occur. In general, r to s odds will be taken to mean the same thing as r/s to 1, i.e., the ratio between the two numbers is the only quantity of importance when stating odds.

Now if it is r times as likely that E will occur as that E will not occur, then the probability that E occurs must be $r/(r+1)$, since we have

$$P(E) = rP(\tilde{E})$$

and

$$P(E) + P(\tilde{E}) = 1.$$

In general, the statement that the odds are r to s in favour of an event E occurring is equivalent to the statement that

$$P(E) = \frac{r/s}{(r/s)+1}$$

$$= \frac{r}{r+s}.$$

If we let $P(E) = p$, then the above equation can easily be solved for r/s in terms of p; we obtain $r/s = p/(1 - p)$. We summarise the above discussion in the following definition.

Definition: If $P(E) = p$, the odds in favour of the event E occurring are $r : s$ (r to s) where $r/s = p/(1 - p)$. If r and s are given, then p can be found by using the equation $p = r/(r + s)$.

Example: Previously we assigned probability 1/5 to the event that candidate C wins the race. Thus, the odds in favour of C winning are 1/5 : 4/5. These odds could equally well have been written as 1 : 4, 2 : 8, and so forth. A bet that C wins is fair if we receive 4 dollars if C wins and pay 1 dollar if C loses.

Infinite Sample Spaces

If a sample space has an infinite number of points, then the way that a distribution function is defined depends upon whether or not the sample space is countable. A sample space is countably infinite if the elements can be counted, i.e., can be put in one-to-one correspondence with the positive integers, and uncountably infinite otherwise. Infinite sample spaces require new concepts in general, but countably infinite spaces do not. If

$$\Omega = \{w_1, w_2, w_3, ...\}$$

is a countably infinite sample space except that the sum must now be a convergent infinite sum. Theorem (a) is still true, as are its extensions Theorems (b) and (d). One thing we cannot do on a countably infinite sample space that we could do on a finite sample space is to define a uniform distribution function. You are asked in Exercise 20 to explain why this is not possible.

Example: A coin is tossed until the first time that a head turns up. Let the outcome of the experiment, w, be the first time that a head turns up. Then the possible outcomes of our experiment are

$$\Omega = \{1, 2, 3, ... \}.$$

Note that even though the coin could come up tails every time we have not allowed for this possibility. We will explain why in a moment. The probability that heads comes up on the first toss is 1/2. The probability that tails comes up on the first toss and heads on the second is 1/4. The probability that we have two tails followed by a head is 1/8, and so forth. This suggests assigning the distribution function $m(n) = 1/2^n$ for $n = 1, 2, 3, \ldots$. To see that this is a distribution function we must show that

$$\sum_{w} m(w) = \frac{1}{2}+\frac{1}{4}+\frac{1}{8}+.... = 1$$

That this is true follows from the formula for the sum of a geometric series,

$$1 + r + r^2 + r^3 + ... = \frac{1}{1-r},$$

or

$$r + r^2 + r^3 + r^4 + ... = \frac{r}{1-r},$$

for $-1 < r < 1$

Putting $r = 1/2$, we see that we have a probability of 1 that the coin eventually turns up heads. The possible outcome of tails every time has to be assigned probability 0, so we omit it from our sample space of possible outcomes.

Let E be the event that the first time a head turns up is after an even number of tosses. Then

$$E = \{2, 4, 6, 8, ... \},$$

and

$$P(E) = \frac{1}{4}+\frac{1}{16}+\frac{1}{64}+\ldots.$$

Putting $r = 1/4$ in $\frac{v}{1-v}$ see that

$$P(E) = \frac{1/4}{1-1/4} = \frac{1}{3}.$$

Thus, the probability that a head turns up for the first time after an even number of tosses is 1/3 and after an odd number of tosses is 2/3.

Historical Remarks

An interesting question in the history of science is: Why was probability not developed until the sixteenth century? We know that in the sixteenth century problems in gambling and games of chance made people start to think about probability. But gambling and games of chance are almost as old as civilization itself. In ancient Egypt (at the time of the First Dynasty, ca. 3500 BC), a game now called "Hounds and Jackals" was played. In this game, the movement of the hounds and jackals was based on the outcome of the roll of four-sided dice made out of animal bones called *astragali*. Six-sided dice made of a variety of materials date back to the sixteenth century BC. Gambling was widespread in ancient Greece and Rome. Indeed, in the Roman Empire it was sometimes found necessary to invoke laws against gambling. Why, then, were probabilities not calculated until the sixteenth century?

Several explanations have been advanced for this late development. One is that the relevant mathematics was not developed and was not easy to develop. The ancient mathematical notation made numerical calculation complicated, and our familiar algebraic notation was not developed until the sixteenth century. However, as we shall see, many of the combinatorial ideas needed to calculate probabilities were discussed long before the sixteenth century. Since many of the chance events of those times had to do with lotteries relating to religious affairs, it has been suggested that there may have been religious barriers to the study of chance and gambling. Another suggestion is that a stronger incentive, such as the development of commerce, was necessary. However, none of these explanations seems completely satisfactory, and people still wonder why it took so long for probability to be studied seriously. An interesting discussion of this problem can be found in Hacking.

The first person to calculate probabilities systematically was Gerolamo Cardano (1501-1576) in his book Liber de Ludo Aleae. This was translated from the Latin by Gould and appears in the book Cardano: The Gambling Scholar by Ore. Ore provides a fascinating discussion of the life of this colourful scholar with accounts of his interests in many different fields, including medicine, astrology, and mathematics. You will also find there a detailed account of Cardano's famous battle with Tartaglia over the solution to the cubic equation.

In his book on probability, Cardano dealt only with the special case that we have called the uniform distribution function. This restriction to equiprobable outcomes was to continue for a long time. In this case Cardano realised that the probability that an event occurs is the ratio of the number of favourable outcomes to the total number of outcomes.

Many of Cardano's examples dealt with rolling dice. Here he realised that the outcomes for two rolls should be taken to be the 36 ordered pairs (i, j) rather than the 21 unordered pairs. This is a subtle point that was still causing problems much later for other writers on probability. For example, in the eighteenth century the famous French mathematician d'Alembert, author of several works on probability, claimed that when a coin is tossed twice the number of heads that turn up would be 0, 1, or 2, and hence we should assign equal probabilities for these three possible outcomes. Cardano chose the correct sample space for his dice problems and calculated the correct probabilities for a variety of events.

Cardano's mathematical work is interspersed with a lot of advice to the potential gambler in short paragraphs, entitled, for example: "Who Should Play and When," "Why Gambling Was Condemned by Aristotle," "Do Those Who Teach Also Play Well?" and so forth. In a paragraph entitled "The Fundamental Principle of Gambling," Cardano writes:

> "The most fundamental principle of all in gambling is simply equal conditions, e.g., of opponents, of bystanders, of money, of situation, of the dice box, and of the die itself.

To the extent to which you depart from that equality, if it is in your opponent's favour, you are a fool, and if in your own, you are unjust."

Cardano did make mistakes, and if he realised it later he did not go back and change his error. For example, for an event that is favourable in three out of four cases, Cardano assigned the correct odds 3 : 1 that the event will occur. But then he assigned odds by squaring these numbers (i.e., 9 : 1) for the event to happen twice in a row. Later, by considering the case where the odds are 1 : 1, he realised that this cannot be correct and was led to the correct result that when f out of n outcomes are favourable, the odds for a favourable outcome twice in a row are $f^2 : n^2 - f^2$. Ore points out that this is equivalent to the realisation that if the probability that an event happens in one experiment is p, the probability that it happens twice is p^2. Cardano proceeded to establish that for three successes the formula should be p^3 and for four successes p^4, making it clear that he understood that the probability is p^n for n successes in n independent repetitions of such an experiment.

Cardano's work was a remarkable first attempt at writing down the laws of probability, but it was not the spark that started a systematic study of the subject. This came from a famous series of letters between Pascal and Fermat. This correspondence was initiated by Pascal to consult Fermat about problems he had been given by Chevalier de Mere, a well-known writer, a prominent figure at the court of Louis XIV, and an ardent gambler.

The first problem de Mere posed was a dice problem. The story goes that he had been betting that at least one six would turn up in four rolls of a die and winning too often, so he then bet that a pair of sixes would turn up in 24 rolls of a pair of dice. The probability of a six with one die is 1/6 and, by the product law for independent experiments, the probability of two sixes when a pair of dice is thrown is $(1/6)(1/6) = 1/36$. Ore claims that a gambling rule of the time suggested that, since four repetitions was favourable for the occurrence of an event with probability 1/6, for an event six times as unlikely, $6 \cdot 4 = 24$ repetitions would be sufficient for a favourable bet. Pascal showed, by exact calculation, that 25 rolls are required for a favourable bet for a pair of sixes.

The second problem was a much harder one: it was an old problem and concerned the determination of a fair division of the stakes in a tournament when the series, for some reason, is interrupted before it is completed. This problem is now referred to as the problem of points. The problem had been a standard problem in mathematical texts; it appeared in Fra Luca Paccioli's book *summa de Arithmetica, Geometria, Proportioni et Proportionalita,* printed in Venice in 1494, in the form:

> "A team plays ball such that a total of 60 points are required to win the game, and each inning counts 10 points. The stakes are 10 ducats. By some incident they cannot finish the game and one side has 50 points and the other 20. One wants to know what share of the prize money belongs to each side. In this case I have found that opinions differ from one to another but all seem to me insufficient in their arguments, but I shall state the truth and give the correct way."

Reasonable solutions, such as dividing the stakes according to the ratio of games won by each player, had been proposed, but no correct solution had been found at the time of the Pascal-Fermat correspondence. The letters deal mainly with the attempts of Pascal and Fermat to solve this problem. Blaise Pascal (1623-1662) was a child prodigy, having published his treatise on conic sections at age sixteen, and having invented a calculating machine at age eighteen. At the time of the letters, his demonstration of the weight of the atmosphere had already established his position at the forefront of contemporary physicists. Pierre de Fermat (1601-1665) was a learned jurist in Toulouse, who studied mathematics in his spare time. He has been called by some the prince of amateurs and one of the greatest pure mathematicians of all times.

The letters, translated by Maxine Merrington, appear in Florence David's fascinating historical account of probability, Games, Gods and Gambling . In a letter dated Wednesday, 29th July, 1654, Pascal writes to Fermat:

"Sir,

Like you, I am equally impatient, and although I am again ill in bed, I cannot help telling you that yesterday evening I received from M. de Carcavi your letter on the problem of points, which I admire more than I can possibly say. I have not the leisure to write at length, but in a word, you have solved the two problems of points, one with dice and the other with sets of games with perfect justness; I am entirely satisfied with it for I do not doubt that I was in the wrong, seeing the admirable agreement in which I find myself with you now. . .

Your method is very sound and is the one which first came to my mind in this research; but because the labour of the combination is excessive, I have found a short cut and indeed another method which is much quicker and neater, which I would like to tell you here in a few words: for henceforth I would like to open my heart to you, if I may, as I am so overjoyed with our agreement. I see that truth is the same in Toulouse as in Paris.

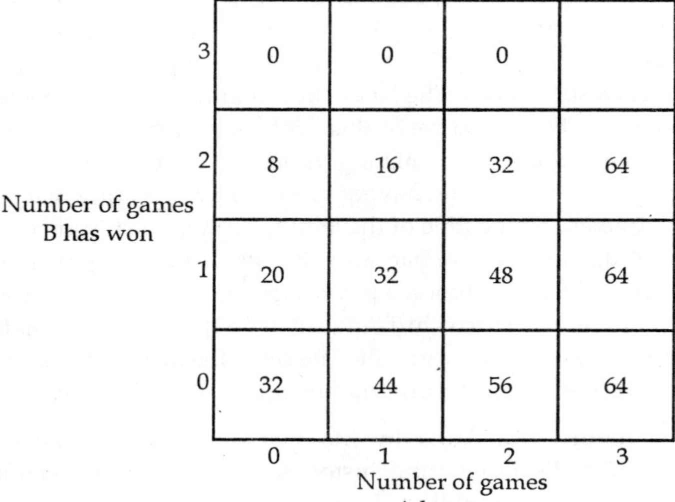

Number of games B has won \ Number of games A has won	0	1	2	3
3	0	0	0	
2	8	16	32	64
1	20	32	48	64
0	32	44	56	64

Figure: Pascal's table.

Here, more or less, is what I do to show the fair value of each game, when two opponents play, for example, in three games and each person has staked 32 pistoles.

Let us say that the first man had won twice and the other once; now they play another game, in which the conditions are that, if the first wins, he takes all the stakes; that is 64 pistoles; if the other wins it, then they have each won two games, and therefore, if they wish to stop playing, they must each take back their own stake, that is, 32 pistoles each.

Then consider, Sir, if the first man wins, he gets 64 pistoles; if he loses he gets 32. Thus if they do not wish to risk this last game but wish to separate without playing it, the first man must say: 'I am certain to get 32 pistoles, even if I lost I still get them; but as for the other 32, perhaps I will get them, perhaps you will get them, the chances are equal. Let us then divide these 32 pistoles in half and give one half to me as well as my 32 which are mine for sure.' He will then have 48 pistoles and the other 16. . ."

Pascal's argument produces the table illustrated in the above figure for the amount due player A at any quitting point.

Each entry in the table is the average of the numbers just above and to the right of the number. This fact, together with the known values when the tournament is completed, determines all the values in this table. If player A wins the first game, then he needs two games to win and B needs three games to win; and so, if the tournament is called off, A should receive 44 pistoles.

The letter in which Fermat presented his solution has been lost; but fortunately, Pascal describes Fermat's method in a letter dated Monday, 24th August, 1654. From Pascal's letter:

"This is your procedure when there are two players: If two players, playing several games, find themselves in that position when the first man needs two games and second needs three, then to find the fair division of stakes,

> you say that one must know in how many games the play will be absolutely decided.
>
> It is easy to calculate that this will be in four games, from which you can conclude that it is necessary to see in how many ways four games can be arranged between two players, and one must see how many combinations would make the first man win and how many the second and to share out the stakes in this proportion. I would have found it difficult to understand this if I had not known it myself already; in fact you had explained it with this idea in mind."

Fermat realised that the number of ways that the game might be finished may not be equally likely. For example, if A needs two more games and B needs three to win, two possible ways that the tournament might go for A to win are WLW and LWLW. These two sequences do not have the same chance of occurring. To avoid this difficulty, Fermat extended the play, adding fictitious plays, so that all the ways that the games might go have the same length, namely four. He was shrewd enough to realise that this extension would not change the winner and that he now could simply count the number of sequences favourable to each player since he had made them all equally likely. If we list all possible ways that the extended game of four plays might go, we obtain the following 16 possible outcomes of the play:

WWWW	WLWW	LWWW	LLWW
WWWL	WLWL	LWWL	LLWL
WWLW	WLLW	LWLW	LLLW
WWLL	WLLL	LWLL	LLLL .

Player *A* wins in the cases where there are at least two wins (the 11 underlined cases), and *B* wins in the cases where there are at least three losses (the other 5 cases). Since A wins in 11 of the 16 possible cases Fermat argued that the probability that A wins is 11/16. If the stakes are 64 pistoles, A should receive 44 pistoles in agreement with Pascal's result. Pascal and Fermat developed more systematic methods for counting the number of favourable outcomes for problems like this, and this will be one of our central problems. Such counting methods fall under the subject of combinatorics.

We see that these two mathematicians arrived at two very different ways to solve the problem of points. Pascal's method was to develop an algorithm and use it to calculate the fair division. This method is easy to implement on a computer and easy to generalise. Fermat's method, on the other hand, was to change the problem into an equivalent problem for which he could use counting or combinatorial methods. In fact, Fermat used what has become known as *Pascal's triangle.* In our study of probability today we shall find that both the algorithmic approach and the combinatorial approach share equal billing, just as they did 300 years ago when probability got its start.

Exercises

1. Let $\Omega = \{a, b, c\}$ be a sample space. Let $m(a) = 1/2$, $m(b) = 1/3$, and $m(c) = 1/6$. Find the probabilities for all eight subsets of Ω.
2. Give a possible sample space Ω for each of the following experiments:
   (a) An election decides between two candidates A and B.
   (b) A two-sided coin is tossed.
   (c) A student is asked for the month of the year and the day of the week on which her birthday falls.
   (d) A student is chosen at random from a class of ten students.
   (e) You receive a grade in this course.
3. For which of the cases in Exercise 2 would it be reasonable to assign the uniform distribution function?
4. Describe in words the events specified by the following subsets of

   Ω = {HHH, HHT, HTH, HTT, THH, THT, TTH, TTT}

   (a) E = {HHH, HHT, HTH, HTT}.
   (b) E = {HHH, TTT}.

(c) E = {HHT, HTH, THH}.

(d) E = {HHT, HTH, HTT, THH, THT, TTH, TTT}.

5. What are the probabilities of the events described in Exercise 4?

6. A die is loaded in such a way that the probability of each face turning up is proportional to the number of dots on that face. (For example, a six is three times as probable as a two). What is the probability of getting an even number in one throw?

7. Let A and B be events such that $P(A \cap B) = 1/4$, $P(\tilde{A}) = 1/3$, and $P(B) = 1/2$. What is $P(A \cup B)$?

8. A student must choose one of the subjects, art, geology, or psychology, as an elective. She is equally likely to choose art or psychology and twice as likely to choose geology. What are the respective probabilities that she chooses art, geology, and psychology?

9. A student must choose exactly two out of three electives: art, French, and mathematics. He chooses art with probability 5/8, French with probability 5/8, and art and French together with probability 1/4. What is the probability that he chooses mathematics? What is the probability that he chooses either art or French?

10. For a bill to come before the President of the United States, it must be passed by both the House of Representatives and the Senate. Assume that, of the bills presented to these two bodies, 60 per cent pass the House, 80 per cent pass the Senate, and 90 per cent pass at least one of the two. Calculate the probability that the next bill presented to the two groups will come before the President.

11. What odds should a person give in favour of the following events?

(a) A card chosen at random from a 52-card deck is an ace.

(b) Two heads will turn up when a coin is tossed twice.

(c) Boxcars (two sixes) will turn up when two dice are rolled.

12. You offer 3 : 1 odds that your friend Smith will be elected mayor of your city. What probability are you assigning to the event that Smith wins?

13. In a horse race, the odds that Romance will win are listed as 2 : 3 and that Downhill will win are 1 : 2. What odds should be given for the event that either Romance or Downhill wins?

14. Let X be a random variable with distribution function $m_X(x)$ defined by

    $m_X(-1) = 1/5$; $m_X(0) = 1/5$, $m_X(1) = 2/5$; $m_X(2) = 1/5$.

    (a) Let Y be the random variable defined by the equation $Y = X + 3$. Find the distribution function $m_Y(y)$ of Y.

    (b) Let Z be the random variable defined by the equation $Z = X^2$. Find the distribution function $m_Z(z)$ of Z.

15. John and Mary are taking a mathematics course. The course has only three grades: A, B, and C. The probability that John gets a B is .3. The probability that Mary gets a B is .4. The probability that neither gets an A but at least one gets a B is .1. What is the probability that at least one gets a B but neither gets a C?

16. In a fierce battle, not less than 70 per cent of the soldiers lost one eye, not less than 75 per cent lost one ear, not less than 80 per cent lost one hand, and not less than 85 per cent lost one leg. What is the minimal possible percentage of those who simultaneously lost one ear, one eye, one hand, and one leg?

17. Assume that the probability of a "success" on a single experiment with n outcomes is $1/n$. Let m be the number of experiments necessary to make it a favourable bet that at least one success will occur.

    (a) Show that the probability that, in m trials, there are no successes is $(1 - 1/n)^m$.

    (b) (de Moivre) Show that if $m = n \log 2$ then

$$\lim_{n\to\infty}\left(1-\frac{1}{n}\right)^m = \frac{1}{2}$$

Hint:

$$\lim_{n\to\infty}\left(1-\frac{1}{n}\right)^n = e^{-1}.$$

Hence, for large n we should choose m to be about $n \log 2$.

(c) Would DeMoivre have been led to the correct answer for de Mere's two bets if he had used his approximation?

18. (a) For events $A_1, \ldots, A_n$, prove that

$$P(A_1 \cup \ldots \cup A_n) \le P(A_1)+\ldots+P(A_n)$$

(b) For events A and B, prove that

$$P(A \cap B) \ge P(A)+P(B)-1$$

19. If A, B, and C are any three events, show that

$$P(A \cup B \cup C) = P(A)+P(B)+P(C)-P(A\cap B)-P(B\cap C)$$
$$-P(C\cap A)+P(A\cap B\cap C)$$

20. Explain why it is not possible to define a uniform distribution function on a countably infinite sample space. *Hint*: Assume $m(w) = a$ for all w, where $0 \le a \le 1$. Does $m(w)$ have all the properties of a distribution function?

21. A die is rolled until the first time that a six turns up. We shall see that the probability that this occurs on the nth roll is $(5/6)^{n-1}.(1/6)$. Using this fact, describe the appropriate infinite sample space and distribution function for the experiment of rolling a die until a six turns up for the first time. Verify that for your distribution function $\sum_w m(w) = 1$.

22. Let Ω be the sample space

$$\Omega = \{0,1,2,\ldots..\},$$

and define a distribution function by

$$m(j) = (1-r)^j r,$$

for some fixed r, $0 < r < 1$, and for $j = 0, 1, 2, \ldots..$ Show that this is a distribution function for Ω.

23. Our calendar has a 400-year cycle. B. H. Brown noticed that the number of times the thirteenth of the month falls on each of the days of the week in the 4,800 months of a cycle is as follows:

Sunday	687
Monday	685
Tuesday	685
Wednesday	687
Thursday	684
Friday	688
Saturday	684

From this he deduced that the thirteenth was more likely to fall on Friday than on any other day. Explain what he meant by this.

24. Tversky and Kahneman asked a group of subjects to carry out the following task. They are told that:

"Linda is 31, single, outspoken, and very bright. She majored in philosophy in college. As a student, she was deeply concerned with racial discrimination and other social issues, and participated in anti-nuclear demonstrations."

The subjects are then asked to rank the likelihood of various alternatives, such as:

(a) Linda is active in the feminist movement.

(b) Linda is a bank teller.

(c) Linda is a bank teller and active in the feminist movement.

Tversky and Kahneman found that between 85 and 90 per cent of the subjects rated alternative (1) most likely, but alternative (3) more likely than alternative (2). Is it? They call this phenomenon the *conjunction fallacy,* and note that it appears to be unaffected by prior training in probability or statistics. Explain why this is a fallacy. Can you give a possible explanation for the subjects' choices?

25. Two cards are drawn successively from a deck of 52 cards. Find the probability that the second card is higher in rank than the first card. *Hint*: Show that 1 = P (higher) + P (lower) + P (same) and use the fact that P (higher) = P (lower).

26. A life table is a table that lists for a given number of births the estimated number of people who will live to a given age.

27. Here is an attempt to get around the fact that we cannot choose a "random integer."

    (a) What, intuitively, is the probability that a "randomly chosen" positive integer is a multiple of 3?

    (b) Let $P_3(N)$ be the probability that an integer, chosen at random between 1 and N, is a multiple of 3 (since the sample space is finite, this is a legitimate probability). Show that the limit

    $$P_3 = \lim_{N \to \infty} P_3(N)$$

    exists and equals 1/3. This formalises the intuition in (a), and gives us a way to assign "probabilities" to certain events that are infinite subsets of the positive integers.

    (c) If A is any set of positive integers, let $A(N)$ mean the number of elements of A which are less than or equal to N. Then define the "probability" of A as

    $$P(A) = \lim_{N \to \infty} A(N)/(N),$$

    provided this limit exists. Show that this definition would assign probability 0 to any finite set and probability 1 to the set of all positive integers. Thus, the probability of the set of all integers is not the sum of the probabilities of the individual integers in this set. This means that the definition of probability given here is not a completely satisfactory definition.

(d) Let A be the set of all positive integers with an odd number of digits. Show that $P(A)$ does not exist. This shows that under the above definition of probability, not all sets have probabilities.

28. (From Sholander) In a standard cloverleaf interchange, there are four ramps for making right-hand turns, and inside these four ramps, there are four more ramps for making left-hand turns. Your car approaches the interchange from the south. A mechanism has been installed so that at each point where there exists a choice of directions, the car turns to the right with fixed probability r.

    (a) If $r = 1{=}2$, what is your chance of emerging from the interchange going west?

    (b) Find the value of r that maximises your chance of a westward departure from the interchange.

29. (From Benkoski) Consider a "pure" cloverleaf interchange in which there are no ramps for right-hand turns, but only the two intersecting straight highways with cloverleaves for left-hand turns. (Thus, to turn right in such an interchange, one must make three left-hand turns). As in the preceding problem, your car approaches the interchange from the south. What is the value of r that maximises your chances of an eastward departure from the interchange?

30. (From vos Savant) A reader of Marilyn vos Savant's column wrote in with the following question:

    "My dad heard this story on the radio. At Duke University, two students had received A's in chemistry all semester. But on the night before the final exam, they were partying in another state and didn't get back to Duke until it was over. Their excuse to the professor was that they had a flat tire, and they asked if they could take a make-up test. The professor agreed, wrote out a test and sent the two to separate rooms to take it. The first question (on one side of the paper) was worth 5 points, and they answered it easily. Then they flipped the paper over and found the second question, worth 95 points: 'Which tire was it?' What was the probability that both students would say the same thing? My dad and I think it's 1 in 16. Is that right?"

(a) Is the answer 1/16?

(b) The following question was asked of a class of students. "I was driving to school today, and one of my tires went flat. Which tire do you think it was?" The responses were as follows: right front, 58 per cent, left front, 11 per cent, right rear, 18 per cent, left rear, 13 per cent. Suppose that this distribution holds in the general population, and assume that the two test-takers are randomly chosen from the general population. What is the probability that they will give the same answer to the second question?

Conditional Probability

Discrete Conditional Probability

Conditional Probability

Suppose we assign a distribution function to a sample space and then learn that an event E has occurred. How should we change the probabilities of the remaining events? We shall call the new probability for an event F the conditional probability of F given E and denote it by $P(F \mid E)$.

Example: An experiment consists of rolling a die once. Let X be the outcome. Let F be the event $\{X = 6\}$, and let E be the event $\{X > 4\}$. We assign the distribution function $m(w) = 1/6$ for $w = 1, 2, \ldots, 6$. Thus, $P(F) = 1/6$. Now suppose that the die is rolled and we are told that the event E has occurred. This leaves only two possible outcomes: 5 and 6. In the absence of any other information, we would still regard these outcomes to be equally likely, so the probability of F becomes $1/2$, making $P(F \mid E) = 1/2$.

Example: In the Life Table, one finds that in a population of 100,000 females, 89.835 per cent can expect to live to age 60, while 57.062 per cent can expect to live to age 80. Given that a woman is 60, what is the probability that she lives to age 80?

This is an example of a conditional probability. In this case, the original sample space can be thought of as a set of 100,000 females. The events E and F are the subsets of the sample space consisting of all women who live at least 60 years, and at least 80 years, respectively. We consider E to be the new sample space, and note that F is a subset of E. Thus, the size of E is 89,835, and the size of F is 57,062. So, the probability in question equals 57,062/ 89,835 = .6352. Thus, a woman who is 60 has a 63.52 per cent chance of living to age 80.

Example: Three candidates A, B, and C are running for office. We decided that A and B have an equal chance of winning and C is only 1/2 as likely to win as A. Let A be the event "A wins," B that "B wins," and C that "C wins." Hence, we assigned probabilities $P(A) = 2/5$, $P(B) = 2/5$, and $P(C) = 1/5$.

Suppose that before the election is held, A drops out of the race. As in previous example, it would be natural to assign new probabilities to the events B and C which are proportional to the original probabilities. Thus, we would have $P(B \mid A) = 2/3$, and $P(C \mid A) = 1/3$. It is important to note that any time we assign probabilities to real-life events, the resulting distribution is only useful if we take into account all relevant information. In this example, we may have knowledge that most voters who favour A will vote for C if A is no longer in the race. This will clearly make the probability that C wins greater than the value of 1/3 that was assigned above.

In these examples we assigned a distribution function and then were given new information that determined a new sample space, consisting of the outcomes that are still possible, and caused us to assign a new distribution function to this space.

We want to make formal the procedure carried out in these examples. Let $\Omega = \{w_1, w_2, \ldots, w_r\}$ be the original sample space with distribution function $m(w_j)$ assigned. Suppose we learn that the event E has occurred. We want to assign a new distribution function $m(w_j \mid E)$ to Ω to reflect this fact. Clearly, if a sample point w_j is not in E, we want $m(w_j \mid E) = 0$. Moreover, in the absence of information to the contrary, it is reasonable to assume that the probabilities for w_k in E should have the same relative magnitudes that they had before we learned that E had occurred.

For this we require that

$$m(w_k \mid E) = cm(w_k)$$

for all w_k in E, with c some positive constant. But we must also have

$$\sum_E m(w_k \mid E) = c\sum_E m(w_k) = 1.$$

Thus,

$$c = \frac{1}{\sum_E m(w_k)} = \frac{1}{P(E)}.$$

(Note that this requires us to assume that $P(E) > 0$). Thus, we will define

$$m(w_k \mid E) = \frac{m(w_k)}{P(E)}$$

for w_k in E. We will call this new distribution the conditional distribution given E. For a general event F, this gives

$$P(F \mid E) = \sum_{F\cap E} m(w_k \mid E) = \sum_{F\cap E} \frac{m(w_k)}{P(E)} = \frac{P(F\cap E)}{P(E)}$$

We call $P(F \mid E)$ the conditional probability of F occurring given that E occurs, and compute it using the formula

$$P(F \mid E) = \frac{P(F\cap E)}{P(E)}.$$

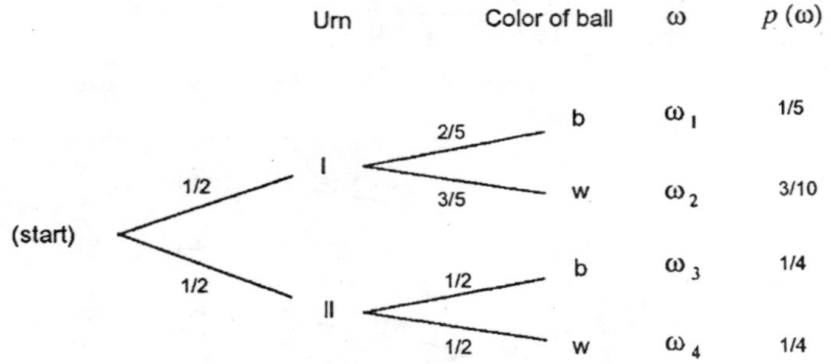

Urn Color of ball ω p (ω)
(start)
1/2
1/2
I
II
2/5
3/5
1/2
1/2
b ω1 1/5
w ω2 3/10
b ω3 1/4
w ω4 1/4

Figure: Tree diagram.

Example: Let us return to die. Recall that F is the event $X = 6$, and E is the event $X > 4$. Note that $E \cap F$ is the event F. So, the above formula gives

$$P(F \mid E) = \frac{P(F \cap E)}{P(E)}$$

$$= \frac{1/6}{1/3}$$

$$= \frac{1}{2},$$

in agreement with the calculations performed earlier.

Example: We have two urns, I and II. Urn I contains 2 black balls and 3 white balls. Urn II contains 1 black ball and 1 white ball. An urn is drawn at random and a ball is chosen at random from it. We can represent the sample space of this experiment as the paths through a tree as shown in the previous figure. The probabilities assigned to the paths are also shown. Let B be the event "a black ball is drawn," and I the event "urn I is chosen." Then the branch weight 2/5, which is shown on one branch in the figure, can now be interpreted as the conditional probability $P(B \mid I)$. Suppose we wish to calculate $P(I \mid B)$. Using the formula, we obtain

$$P(I \mid B) = \frac{P(I \cap B)}{P(B)}$$

$$= \frac{P(I \cap B)}{P(B \cap I) + P(B \cap II)}$$

$$= \frac{1/5}{1/5 + 1/4} = \frac{4}{9}.$$

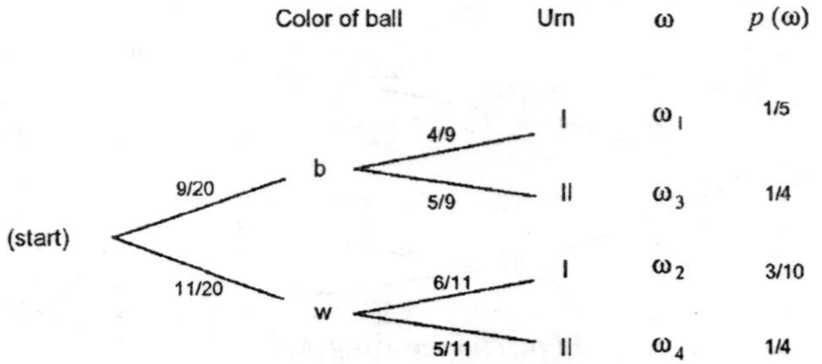

Figure: Reverse tree diagram.

Bayes Probabilities

Our original tree measure gave us the probabilities for drawing a ball of a given colour, given the urn chosen. We have just calculated the inverse probability that a particular urn was chosen, given the colour of the ball. Such an inverse probability is called a Bayes probability and may be obtained by a formula that we shall develop later. Bayes probabilities can also be obtained by simply constructing the tree measure for the two-stage experiment carried out in reverse order. We show this tree in the previous figure.

The paths through the reverse tree are in one-to-one correspondence with those in the forward tree, since they correspond to individual outcomes of the experiment, and so they are assigned the same probabilities. From the forward tree, we find that the probability of a black ball is

$$\frac{1}{2}\cdot\frac{2}{5}+\frac{1}{2}\cdot\frac{1}{2}=\frac{9}{20}.$$

The probabilities for the branches at the second level are found by simple division. For example, if x is the probability to be assigned to the top branch at the second level, we must have

$$\frac{9}{20}\cdot x = \frac{1}{5}$$

or $x = 4/9$. Thus, $P(I \mid B) = 4/9$, in agreement with our previous calculations. The reverse tree then displays all of the inverse, or Bayes, probabilities.

Example: We consider now a problem called the Monty Hall problem. This has long been a favourite problem but was revived by a letter from Craig Whitaker to Marilyn vos Savant for consideration in her column in Parade Magazine. Craig wrote:

> Suppose you're on Monty's Hall. Let's Make a Deal! You are given the choice of three doors, behind one door is a car, the others, goats. You pick a door, say 1, Monty opens another door, say 3, which has a goat. Monty says to you "Do you want to pick door 2?" Is it to your advantage to switch your choice of doors?

Marilyn gave a solution concluding that you should switch, and if you do, your probability of winning is 2/3. Several irate readers, some of whom identified themselves as having a Ph.D. in mathematics, said that this is absurd since after Monty has ruled out one door, there are only two possible doors and they should still each have the same probability 1/2 so there is no advantage to switching. Marilyn stuck to her solution and encouraged her readers to simulate the game and draw their own conclusions from this. We also encourage the reader to do this.

Other readers complained that Marilyn had not described the problem completely. In particular, the way in which certain decisions were made during a play of the game were not specified. We will assume that the car was put behind a door by rolling a three-sided die which made all three choices equally likely. Monty knows where the car is, and always opens a door with a goat behind it. Finally, we assume that if Monty has a choice of doors (i.e., the contestant has picked the door with the car behind it), he chooses each door with probability 1/2. Marilyn clearly expected her readers to assume that the game was played in this manner.

As is the case with most apparent paradoxes, this one can be resolved through careful analysis. We begin by describing a simpler, related question. We say that a contestant is using the "stay" strategy if he picks a door, and, if offered a chance to switch to another door, declines to do so (i.e., he stays with his original choice).

Similarly, we say that the contestant is using the "switch" strategy if he picks a door, and, if offered a chance to switch to another door, takes the offer. Now suppose that a contestant decides in advance to play the "stay" strategy. His only action in this case is to pick a door (and decline an invitation to switch, if one is offered). What is the probability that he wins a car? The same question can be asked about the "switch" strategy.

Using the "stay" strategy, a contestant will win the car with probability 1/3, since 1/3 of the time the door he picks will have the car behind it. On the other hand, if a contestant plays the "switch" strategy, then he will win whenever the door he originally picked does not have the car behind it, which happens 2/3 of the time.

This very simple analysis, though correct, does not quite solve the problem that Craig posed. Craig asked for the conditional probability that you win if you switch, given that you have chosen door 1 and that Monty has chosen door 3. To solve this problem, we set up the problem before getting this information and then compute the conditional probability given this information. This is a process that takes place in several stages; the car is put behind a door, the contestant picks a door, and finally Monty opens a door. Thus it is natural to analyse this using a tree measure. Here we make an additional assumption that if Monty has a choice of doors (i.e., the contestant has picked the door with the car behind it) then he picks each door with probability 1/2. The assumptions we have made determine the branch probabilities and these in turn determine the tree measure. The resulting tree and tree measure are shown in the figure below. It is tempting to reduce the tree's size by making certain assumptions such as: "Without loss of generality, we will assume that the contestant always picks door 1." We have chosen not to make any such assumptions, in the interest of clarity.

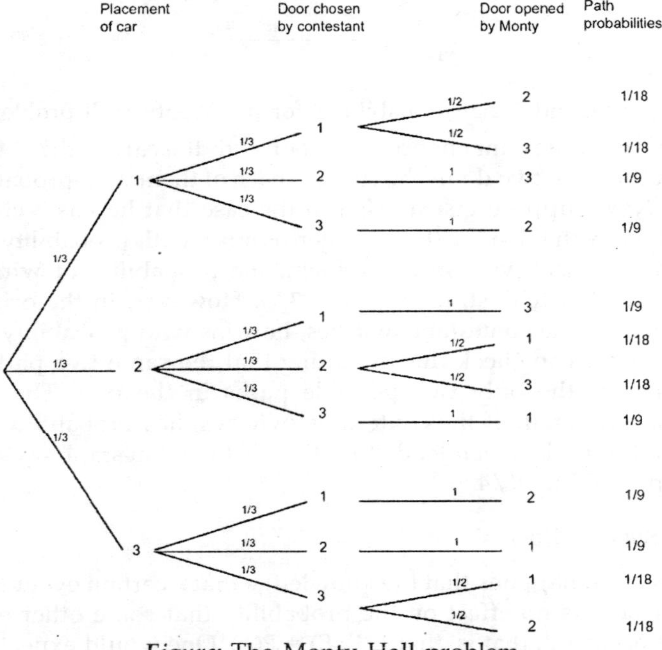

Figure: The Monty Hall problem.

Now the given information, namely that the contestant chose door 1 and Monty chose door 3, means only two paths through the tree are possible. For one of these paths, the car is behind door 1 and for the other it is behind door 2. The path with the car behind door 2 is twice as likely as the one with the car behind door 1. Thus the conditional probability is 2/3 that the car is behind door 2 and 1/3 that it is behind door 1, so if you switch you have a 2/3 chance of winning the car, as Marilyn claimed.

At this point, the reader may think that the two problems above are the same, since they have the same answers. Recall that we assumed in the original problem

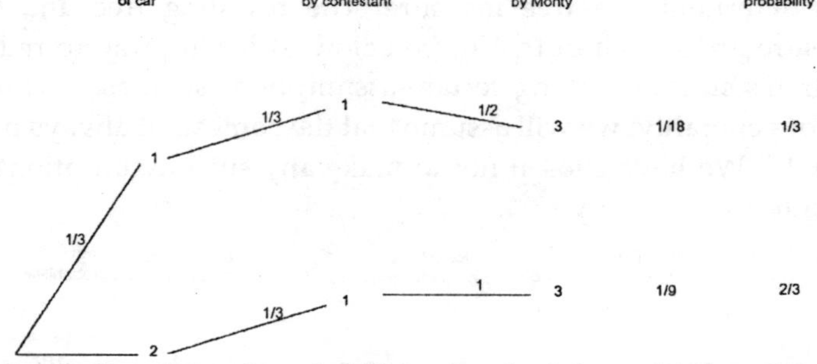

Figure: Conditional probabilities for the Monty Hall problem.

if the contestant chooses the door with the car, so that Monty has a choice of two doors, he chooses each of them with probability 1/2. Now suppose instead that in the case that he has a choice, he chooses the door with the larger number with probability 3/4. In the "switch" vs. "stay" problem, the probability of winning with the "switch" strategy is still 2/3. However, in the original problem, if the contestant switches, he wins with probability 4/7. The reader can check this by noting that the same two paths as before are the only two possible paths in the tree. The path leading to a win, if the contestant switches, has probability 1/3, while the path which leads to a loss, if the contestant switches, has probability 1/4.

Independent Events

It often happens that the knowledge that a certain event E has occurred has no effect on the probability that some other event F has occurred, that is, that $P(F \mid E) = P(F)$. One would expect that in this case, the equation $P(E \mid F) = P(E)$ would also be true. In fact, each equation implies the other. If these equations are true, we might say the F is independent of E. For example, you would not expect the knowledge of the outcome of the first toss of a coin to change the probability that you would assign to the possible outcomes of the second toss, that is, you would not expect that the second toss depends on the first. This idea is formalised in the following definition of independent events.

Definition: Two events E and F are independent if both E and F have positive probability and if

$$P(E \mid F) = P(E),$$

and

$$P(F \mid E) = P(F).$$

As noted above, if both $P(E)$ and $P(F)$ are positive, then each of the above equations imply the other, so that to see whether two events are independent, only one of these equations must be checked.

The following theorem provides another way to check for independence.

Theorem: If $P(E) > 0$ and $P(F) > 0$, then E and F are independent if and only if

$$P(E \cap F) = P(E)P(F).$$

Proof: Assume first that E and F are independent. Then $P(E \mid F) = P(E)$, and so

$$P(E \cap F) = P(E \mid F)P(F)$$

$$= P(E)P(F).$$

Assume next that $P(E \cap F) = P(E)P(F)$. Then

$$P(E \mid F) = \frac{P(E \cap F)}{P(F)} = P(E).$$

Also,

$$P(F \mid E) = \frac{P(F \cap E)}{P(E)} = P(F).$$

Therefore, E and F are independent.

Example: Suppose that we have a coin which comes up heads with probability p, and tails with probability q. Now suppose that this coin is tossed twice. Using a frequency interpretation of probability, it is reasonable to assign to the outcome (H, H) the probability p^2, to the outcome (H, T) the probability pq, and so on. Let E be the event that heads turns up on the first toss and F the event that tails turns up on the second toss. We will now check that with the above probability assignments, these two events are independent, as expected. We have $P(E) = p^2 + pq = p$, $P(F) = pq + q^2 = q$. Finally $P(E \cap F) = pq$, so $P(E \cap F) = P(E)P(F)$.

Example: It is often, but not always, intuitively clear when two events are independent. In the previous example, let A be the event "the first toss is a head" and B the event "the two outcomes are the same." Then

$$P(B \mid A) = \frac{P(B \cap A)}{P(A)} = \frac{P\{HH\}}{P\{HH, HT\}} = \frac{1/4}{1/2} = \frac{1}{2} = P(B).$$

Therefore, A and B are independent, but the result was not so obvious.

Example: Finally, let us give an example of two events that are not independent. In Example 4.7, let I be the event "heads on the first toss" and J the event "two heads turn up." Then $P(I) = 1/2$ and $P(J) = 1/4$. The event $I \cap J$ is the event "heads on both tosses" and has probability 1/4. Thus, I and J are not independent since $P(I)P(J) = 1/8 \neq P(I \cap J)$.

We can extend the concept of independence to any finite set of events $A_1, A_2, \ldots, A_n$.

Definition: A set of events $\{A_1, A_2, \ldots, A_n\}$ is said to be mutually independent if for any subset $\{A_i, A_j, \ldots, A_m\}$ of these events we have

$$P(A_i \cap A_j \cap \ldots \cap A_m) = P(A_i)P(A_j)\ldots P(A_m),$$

or equivalently, if for any sequence $\bar{A}_1, \bar{A}_2, \ldots, \bar{A}_n$ with $\bar{A}_j =$ Aj or $\tilde{A}_j$,

$$P(\bar{A}_1 \cap \bar{A}_2 \cap \ldots \cap \bar{A}_n) = P(\bar{A}_1)P(\bar{A}_2)\ldots P(\bar{A}_n).$$

Using this terminology, it is a fact that any sequence (*S, S, F, F, S, . . ., S*) of possible outcomes of a Bernoulli trials process forms a sequence of mutually independent events.

It is natural to ask: If all pairs of a set of events are independent, is the whole set mutually independent?

It is important to note that the statement

$$P(A_1 \cap A_2 \cap ... \cap A_n) = P(A_1)P(A_2)...P(A_n)$$

does not imply that the events $A_1, A_2, . . ., A_n$ are mutually independent.

Joint Distribution Functions and Independence of Random Variables

It is frequently the case that when an experiment is performed, several different quantities concerning the outcomes are investigated.

Example: Suppose we toss a coin three times. The basic random variable $\overline{X}$ corresponding to this experiment has eight possible outcomes, which are the ordered triples consisting of H's and T's. We can also define the random variable X_i, for $i = 1, 2, 3$, to be the outcome of the *i*th toss. If the coin is fair, then we should assign the probability 1/8 to each of the eight possible outcomes. Thus, the distribution functions of X_1, X_2, and X_3 are identical; in each case they are defined by $m(H) = m(T) = 1/2$.

If we have several random variables $X_1, X_2, . . ., X_n$ which correspond to a given experiment, then we can consider the joint random variable $\overline{X} = (X_1, X_2, . . ., X_n)$ defined by taking an outcome *w* of the experiment, and writing, as an *n*-tuple, the corresponding *n* outcomes for the random variables $X_1, X_2, . . ., X_n$. Thus, if the random variable X_i has, as its set of possible outcomes the set R_i, then the set of possible outcomes of the joint random variable $\overline{X}$ is the Cartesian product of the R_i's, i.e., the set of all *n*-tuples of possible outcomes of the X_i's.

Example: In the coin-tossing example above, let X_i denote the outcome of the *i*th toss. Then the joint random variable $\overline{X} = (X_1, X_2, X_3)$ has eight possible outcomes.

Suppose that we now define Y_i, for $i = 1, 2, 3$, as the number of heads which occur in the first i tosses. Then Y_i has $\{0, 1, ..., i\}$ as possible outcomes, so at first glance, the set of possible outcomes of the joint random variable $\bar{Y} = (Y_1, Y_2, Y_3)$ should be the set

$$\{(a_1, a_2, a_3) : 0 \le a_1 \le 1, 0 \le a_2 \le 2, 0 \le a_3 \le 3\}.$$

However, the outcome (1, 0, 1) cannot occur, since we must have $a_1 \le a_2 \le a_3$. The solution to this problem is to define the probability of the outcome (1, 0, 1) to be 0.

We now illustrate the assignment of probabilities to the various outcomes for the joint random variables $\bar{X}$ and $\bar{Y}$. In the first case, each of the eight outcomes should be assigned the probability 1/8, since we are assuming that we have a fair coin. In the second case, since Y_i has $i + 1$ possible outcomes, the set of possible outcomes has size 24. Only eight of these 24 outcomes can actually occur, namely the ones satisfying $a_1 \le a_2 \le a_3$. Each of these outcomes corresponds to exactly one of the outcomes of the random variable $\bar{X}$, so it is natural to assign probability 1/8 to each of these. We assign probability 0 to the other 16 outcomes. In each case, the probability function is called a joint distribution function.

We collect the above ideas in a definition.

Definition: Let $X_1, X_2, \ldots, X_n$ be random variables associated with an experiment. Suppose that the sample space (i.e., the set of possible outcomes) of X_i is the set R_i. Then the joint random variable $X = (X_1, X_2, \ldots, X_n)$ is defined to be the random variable whose outcomes consist of ordered n-tuples of outcomes, with the ith coordinate lying in the set R_i. The sample space Ω of $\bar{X}$ is the Cartesian product of the R_i's:

$$\Omega = R_1 \times R_2 \times \cdots \times R_n.$$

The joint distribution function of $\bar{X}$ is the function which gives the probability of each of the outcomes of $\bar{X}$.

Example: We now consider the assignment of probabilities in the above example. In the case of the random variable $\bar{X}$, the probability of any outcome (a_1, a_2, a_3) is just the product of the probabilities $P(X_i = a_i)$,

Table: Smoking and Cancer.

	Not smoke	Smoke	Total
Not cancer	40	10	50
Cancer	7	3	10
Totals	47	13	60

Table: Joint Distribution.

		S	
		0	1
C	0	40/60	10/60
	1	7/60	3/60

for $i = 1, 2, 3$. However, in the case of $\bar{Y}$, the probability assigned to the outcome (1, 1, 0) is not the product of the probabilities $P(Y_1 = 1)$, $P(Y_2 = 1)$, and $P(Y_3 = 0)$. The difference between these two situations is that the value of X_i does not affect the value of X_j, if $i \neq j$, while the values of Y_i and Y_j affect one another. For example, if $Y_1 = 1$, then Y_2 cannot equal 0. This prompts the next definition.

Definition: The random variables $X_1, X_2, \ldots, X_n$ are mutually independent if

$$P(X_1 = r_1, X_2 = r_2, \ldots, X_n = r_n) = P(X_1 = r_1)P(X_2 = r_2)\cdots P(X_n = r_n)$$

for any choice of $r_1, r_2, \ldots, r_n$. Thus, if $X_1, X_2, \ldots, X_n$ are mutually independent, then the joint distribution function of the random variable

$$\bar{X} = (X_1, X_2, \ldots, X_n)$$

is just the product of the individual distribution functions. When two random variables are mutually independent, we shall say more briefly that they are independent.

Example: In a group of 60 people, the numbers who do or do not smoke and do or do not have cancer are reported as shown in the previous table. Let Ω be the sample space consisting of these

60 people. A person is chosen at random from the group. Let $C(w) = 1$ if this person has cancer and 0 if not, and $S(w) = 1$ if this person smokes and 0 if not. Then the joint distribution of $\{C, S\}$ is given in the previous table. For example $P(C = 0, S = 0) = 40/60$, $P(C = 0, S = 1) = 10/60$, and so forth. The distributions of the individual random variables are called *marginal distributions.* The marginal distributions of C and S are:

$$p_C = \begin{pmatrix} 0 & 1 \\ 50/60 & 10/60 \end{pmatrix},$$

$$p_S = \begin{pmatrix} 0 & 1 \\ 47/60 & 13/60 \end{pmatrix}.$$

The random variables S and C are not independent, since

$$P(C = 1, S = 1) = \frac{3}{60} = .05\,,$$

$$P(C = 1)P(S = 1) = \frac{10}{60} \cdot \frac{13}{60} = .036.$$

Note that we would also see this from the fact that

$$P(C = 1 \mid S = 1) = \frac{3}{13} = .23,$$

$$P(C = 1) = \frac{1}{6} = .167.$$

Independent Trials Processes

The study of random variables proceeds by considering special classes of random variables. One such class that we shall study is the class of independent trials.

Definition: A sequence of random variables $X_1, X_2, \ldots, X_n$ that are mutually independent and that have the same distribution is called a sequence of independent trials or an *independent trials process.*

Independent trials processes arise naturally in the following way. We have a single experiment with sample space $R = \{r_1, r_2, \ldots, r_s\}$ and a distribution function

$$m_X = \begin{pmatrix} r_1 & r_2 & \cdots & r_s \\ p_1 & p_2 & \cdots & p_s \end{pmatrix}.$$

We repeat this experiment n times. To describe this total experiment, we choose as sample space the space

$$\Omega = R \times R \times \cdots \times R,$$

consisting of all possible sequences $w = (w_1, w_2, \ldots, w_n)$ where the value of each w_j is chosen from R. We assign a distribution function to be the product distribution

$$m(w) = m(w_1) \cdot \ldots \cdot m(w_n),$$

with $m(w_j) = p_k$ when $w_j = r_k$. Then we let X_j denote the jth coordinate of the outcome $(r_1, r_2, \ldots, r_n)$. The random variables $X_1, \ldots, X_n$ form an independent trials process.

Example: An experiment consists of rolling a die three times. Let X_i represent the outcome of the ith roll, for $i = 1, 2, 3$. The common distribution function is

$$m_i = \begin{pmatrix} 1 & 2 & 3 & 4 & 5 & 6 \\ 1/6 & 1/6 & 1/16 & 1/6 & 1/6 & 1/6 \end{pmatrix}.$$

The sample space is $R^3 = R \times R \times R$ with $R = \{1, 2, 3, 4, 5, 6\}$. If $w = (1, 3, 6)$, then $X_1(w) = 1$, $X_2(w) = 3$, and $X_3(w) = 6$ indicating that the first roll was a 1, the second was a 3, and the third was a 6. The probability assigned to any sample point is

$$m(w) = \frac{1}{6} \cdot \frac{1}{6} \cdot \frac{1}{6} = \frac{1}{216}.$$

Example: Consider next a Bernoulli trials process with probability p for success on each experiment. Let $X_j(w) = 1$ if the jth outcome is success and $X_j(w) = 0$ if it is a failure. Then $X_1, X_2, \ldots, X_n$ is an independent trials process. Each X_j has the same distribution function

$$m_j = \begin{pmatrix} 0 & 1 \\ q & p \end{pmatrix},$$

where $q = 1 - p$.

If $S_n = X_1 + X_2 + \ldots + X_n$, then

$$P(S_n = j) = \binom{n}{j} p^j q^{n-j},$$

and S_n has, as distribution, the binomial distribution $b(n, p, j)$.

Bayes' Formula

In our examples, we have considered conditional probabilities of the following form: Given the outcome of the second stage of a two-stage experiment, find the probability for an outcome at the first stage. We have remarked that these probabilities are called *Bayes probabilities.*

We return now to the calculation of more general Bayes probabilities. Suppose we have a set of events $H_1, H_2, \ldots, H_m$ that are pairwise disjoint and such that

$$\Omega = H_1 \cup H_2 \cup \cdots \cup H_m.$$

We call these events hypotheses. We also have an event E that gives us some information about which hypothesis is correct. We call this event evidence.

Before we receive the evidence, then, we have a set of prior probabilities $P(H_1), P(H_2), \ldots, P(H_m)$ for the hypotheses. If we know the correct hypothesis, we know the probability for the evidence. That is, we know $P(E \mid H_i)$ for all i. We want to find the probabilities for the hypotheses given the evidence. That is, we want to find the conditional probabilities $P(H_i \mid E)$. These probabilities are called the posterior probabilities.

To find these probabilities, we write them in the form

$$P(H_i \backslash E) = \frac{P(H_i \cap E)}{P(E)}. \tag{4.1}$$

Table: Diseases data.

Disease	Number having this disease	The results + +	+ −	− +	− −
d_1	3215	2110	301	704	100
d_2	2125	396	132	1187	410
d_3	4660	510	3568	73	509
Total	10000				

We can calculate the numerator from our given information by

$$P(H_i \cap E) = P(H_i)P(E/H_i). \tag{4.2}$$

Since one and only one of the events $H_1, H_2, \ldots, H_m$ can occur, we can write the probability of E as

$$P(E) = P(H_1 \cap E) + P(H_2 \cap E) + \cdots + P(H_m \cap E).$$

Using equation 4.2, the above expression can be seen to equal

$$P(H_1)P(E/H_1) + P(H_2)P(E/H_2) + \cdots + P(H_m)P(E/H_m). \tag{4.3}$$

Using (4.1), (4.2), and (4.3) yields *Bayes' formula*:

$$P(H_i/E) = \frac{P(H_i)P(E/H_i)}{\Sigma_{k=1}^{m} P(H_k)P(E/H_k)}.$$

Although this is a very famous formula, we will rarely use it. If the number of hypotheses is small, a simple tree measure calculation is easily carried out, as we have done in our examples. If the number of hypotheses is large, then we should use a computer.

Bayes probabilities are particularly appropriate for medical diagnosis. A doctor is anxious to know which of several diseases a patient might have. She collects evidence in the form of the outcomes of certain tests. From statistical studies the doctor can find the prior probabilities of the various diseases before the tests, and the probabilities for specific test outcomes, given a particular disease. What the doctor wants to know is the posterior probability for the particular disease, given the outcomes of the tests.

Example: A doctor is trying to decide if a patient has one of three diseases *d*1, *d*2, or *d*3. Two tests are to be carried out, each of which results in a positive (+) or a negative (-) outcome. There are four possible test patterns ++, +–, –+, and ––. National records have indicated that, for 10,000 people having one of these three diseases, the distribution of diseases and test results are as in above table.

From this data, we can estimate the prior probabilities for each of the diseases and, given a particular disease, the probability of a particular test outcome. For example, the prior probability of disease d_1 may be estimated to be 3215/10,000 = .3215. The probability of the test result +–, given disease d_1, may be estimated to be 301/3125 = .094.

Table: Posterior probabilities.

	d1	d2	d3
+ +	.700	.132	.168
+ –	.076	.033	.891
– +	.357	.605	.038
– –	.098	.405	.497

We can now use Bayes' formula to compute various posterior probabilities. The computer programme Bayes computes these posterior probabilities. The results for this example are shown in the previous table.

We note from the outcomes that, when the test result is ++, the disease d_1 has a significantly higher probability than the other two. When the outcome is + –, this is true for disease d_3. When the outcome is – +, this is true for disease d_2 .

Note that these statements might have been guessed by looking at the data. If the outcome is – –, the most probable cause is d_3, but the probability that a patient has d_2 is only slightly smaller. If one looks at the data in this case, one can see that it might be hard to guess which of the two diseases d_2 and d_3 is more likely.

Our final example shows that one has to be careful when the prior probabilities are small.

Example: A doctor gives a patient a test for a particular cancer. Before the results of the test, the only evidence the doctor has to go on is that 1 woman in 1000 has this cancer. Experience has shown that, in 99 per cent of the cases in which cancer is present, the test is positive; and in 95 per cent of the cases in which it is not present, it is negative. If the test turns out to be positive, what probability should the doctor assign to the event that cancer is present? An alternative form of this question is to ask for the relative frequencies of false positives and cancers.

We are given that prior(cancer) = .001 and prior(not cancer) = .999. We know also that $P(+ \mid \text{cancer}) = .99$, $P(- \mid \text{cancer}) = .01$, $P(+ \mid \text{not cancer}) = .05$, and $P(- \mid \text{not cancer}) = .95$. Using this data gives the result shown in the figure ahead.

We see now that the probability of cancer given a positive test has only increased from .001 to .019. While this is nearly a twentyfold increase, the probability that the patient has the cancer is still small. Stated in another way, among the positive results, 98.1 per cent are false positives, and 1.9 per cent are cancers. When a group of second-year medical students was asked this question, over half of the students incorrectly guessed the probability to be greater than .5.

Historical Remarks

Conditional probability was used long before it was formally defined. Pascal and Fermat considered the problem of points: given that team A has won m games and team B has won n games, what is the probability that A will win the series? This is clearly a conditional probability problem.

In his book, Huygens gave a number of problems, one of which was:

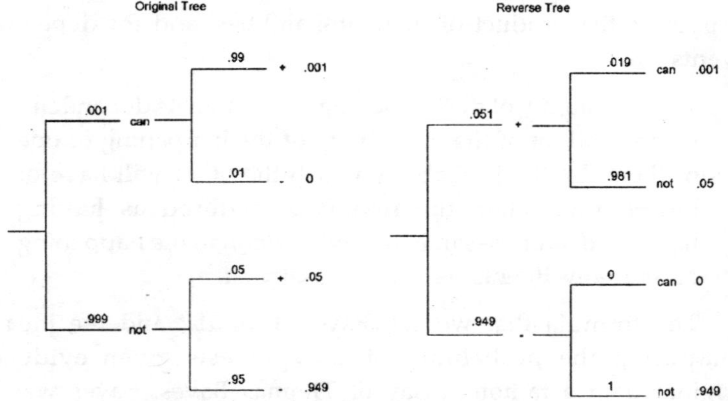

Figure: Forward and reverse tree diagrams.

Three gamblers, A, B and C, take 12 balls of which 4 are white and 8 black. They play with the rules that the drawer is blindfolded, A is to draw first, then B and then C, the winner to be the one who first draws a white ball. What is the ratio of their chances?

From his answer it is clear that Huygens meant that each ball is replaced after drawing. However, John Hudde, the mayor of Amsterdam, assumed that he meant to sample without replacement and corresponded with Huygens about the difference in their answers. Hacking remarks that "Neither party can understand what the other is doing."

By the time of de Moivre's book, *The Doctrine of Chances*, these distinctions were well understood. De Moivre defined independence and dependence as follows:

> Two Events are independent, when they have no connection one with the other, and that the happening of one neither forwards nor obstructs the happening of the other.
>
> Two Events are dependent, when they are so connected together as that the Probability of either's happening is altered by the happening of the other.

De Moivre used sampling with and without replacement to illustrate that the probability that two independent events both happen is the product of their probabilities, and for dependent events that:

> The Probability of the happening of two Events dependent, is the product of the Probability of the happening of one of them, by the Probability which the other will have of happening, when the first is considered as having happened; and the same Rule will extend to the happening of as many Events as may be assigned.

The formula that we call Bayes' formula, and the idea of computing the probability of a hypothesis given evidence, originated in a famous essay of Thomas Bayes. Bayes was an ordained minister in Tunbridge Wells near London. His mathematical interests led him to be elected to the Royal Society in 1742, but none of his results were published within his lifetime. The work upon which his fame rests, "An Essay Towards Solving a Problem in the Doctrine of Chances," was published in 1763, three years after his death. Bayes reviewed some of the basic concepts of probability and then considered a new kind of inverse probability problem requiring the use of conditional probability.

Bernoulli, in his study of processes that we now call Bernoulli trials, had proven his famous law of large numbers. This theorem assured the experimenter that if he knew the probability p for success, he could predict that the proportion of successes would approach this value as he increased the number of experiments. Bernoulli himself realised that in most interesting cases you do not know the value of p and saw his theorem as an important step in showing that you could determine p by experimentation.

To study this problem further, Bayes started by assuming that the probability p for success is itself determined by a random experiment. He assumed in fact that this experiment was such that this value for p is equally likely to be any value between 0 and 1. Without knowing this value we carry out n experiments and observe m successes. Bayes proposed the problem of finding the conditional probability that the unknown probability p lies between a and b. He obtained the answer:

$$P(a \leq p < \mathrm{b} \mid \mathrm{m} \text{ successes in } n \text{ trials}) = \frac{\int_a^b x^m (1-x)^{n-m}\, dx}{\int_{0.}^1 x^m (1-x)^{n-m}\, dx}.$$

Bayes clearly wanted to show that the conditional distribution function, given the outcomes of more and more experiments, becomes concentrated around the true value of p. Thus, Bayes was trying to solve an inverse problem. The computation of the integrals was too difficult for exact solution except for small values of j and n, and so Bayes tried approximate methods. His methods were not very satisfactory and it has been suggested that this discouraged him from publishing his results.

However, his paper was the first in a series of important studies carried out by Laplace, Gauss, and other great mathematicians to solve inverse problems. They studied this problem in terms of errors in measurements in astronomy. If an astronomer were to know the true value of a distance and the nature of the random errors caused by his measuring device he could predict the probabilistic nature of his measurements. In fact, however, he is presented with the inverse problem of knowing the nature of the random errors, and the values of the measurements, and wanting to make inferences about the unknown true value.

As Maistrov remarks, the formula that we have called Bayes' formula does not appear in his essay. Laplace gave it this name when he studied these inverse problems. The computation of inverse probabilities is fundamental to statistics and has led to an important branch of statistics called Bayesian analysis, assuring Bayes eternal fame for his brief essay.

Exercises

1. Assume that E and F are two events with positive probabilities. Show that if $P(E/F) = P(E)$, then $P(F/E) = P(F)$.

2. A coin is tossed three times. What is the probability that exactly two heads occur, given that
   (a) the first outcome was a head?
   (b) the first outcome was a tail?
   (c) the first two outcomes were heads?
   (d) the first two outcomes were tails?
   (e) the first outcome was a head and the third outcome was a head?

3. A die is rolled twice. What is the probability that the sum of the faces is greater than 7, given that
   (a) the first outcome was a 4?
   (b) the first outcome was greater than 3?
   (c) the first outcome was a 1?
   (d) the first outcome was less than 5?

4. A card is drawn at random from a deck of cards. What is the probability that
   (a) it is a heart, given that it is red?
   (b) it is higher than a 10, given that it is a heart? (Interpret J, Q, K, A as 11,12,13,14).
   (c) it is a jack, given that it is red?
5. A coin is tossed three times. Consider the following events
   A: Heads on the first toss.
   B: Tails on the second.
   C: Heads on the third toss.
   D: All three outcomes the same (HHH or TTT).
   E: Exactly one head turns up.
   (a) Which of the following pairs of these events are independent?
      (1) A, B
      (2) A, D
      (3) A, E
      (4) D, E
   (b) Which of the following triples of these events are independent?
      (1) A, B, C
      (2) A, B, D
      (3) C, D, E
6. From a deck of five cards numbered 2, 4, 6, 8, and 10, respectively, a card is drawn at random and replaced. This is done three times. What is the probability that the card numbered 2 was drawn exactly two times, given that the sum of the numbers on the three draws is 12?
7. A coin is tossed twice. Consider the following events.
   A: Heads on the first toss.
   B: Heads on the second toss.
   C: The two tosses come out the same.

a. Show that A, B, C are pairwise independent but not independent.

(b) Show that C is independent of A and B but not of $A \cap B$.

8. Let $\Omega = \{a, b, c, d, e, f\}$. Assume that $m(a) = m(b) = 1/8$ and $m(c) = m(d) = m(e) = m(f) = 3/16$. Let A, B, and C be the events $A = \{d, e, a\}$, $B = \{c, e, a\}$, $C = \{c, d, a\}$. Show that $P(A \cap B \cap C) = P(A)P(B)P(C)$ but no two of these events are independent.

9. What is the probability that a family of two children has

(a) two boys given that it has at least one boy?

(b) two boys given that the first child is a boy?

10. In an example, we used the Life Table to compute a conditional probability. The number 93,753 in the table, corresponding to 40-year-old males, means that of all the males born in the United States in 1950, 93.753 per cent were alive in 1990. Is it reasonable to use this as an estimate for the probability of a male, born this year, surviving to age 40?

11. Simulate the Monty Hall problem. Carefully state any assumptions that you have made when writing the programme. Which version of the problem do you think that you are simulating?

12. In an example, how large must the prior probability of cancer be to give a posterior probability of .5 for cancer given a positive test?

13. Two cards are drawn from a bridge deck. What is the probability that the second card drawn is red?

14. If $P(\tilde{B}) = 1/4$ and $P(A/B) = 1/2$, what is $P(A \cap B)$?

15. (a) What is the probability that your bridge partner has exactly two aces, given that she has at least one ace?

(b) What is the probability that your bridge partner has exactly two aces, given that she has the ace of spades?

16. Prove that for any three events A, B, C, each having positive probability,

$$P(A \cap B \cap C) = P(A)P(B/A)P(C/A \cap B).$$

17. Prove that if A and B are independent so are
    (a) A and $\tilde{B}$.
    (b) $\tilde{A}$ and $\tilde{B}$.

18. A doctor assumes that a patient has one of three diseases d_1, d_2, or d_3. Before any test, he assumes an equal probability for each disease. He carries out a test that will be positive with probability .8 if the patient has d_1, .6 if he has disease d_2, and .4 if he has disease d_3. Given that the outcome of the test was positive, what probabilities should the doctor now assign to the three possible diseases?

19. In a poker hand, John has a very strong hand and bets 5 dollars. The probability that Mary has a better hand is .04. If Mary had a better hand she would raise with probability .9, but with a poorer hand she would only raise with probability .1. If Mary raises, what is the probability that she has a better hand than John does?

20. The Polya urn model for contagion is as follows: We start with an urn which contains one white ball and one black ball. At each second we choose a ball at random from the urn and replace this ball and add one more of the colour chosen. Write a programme to simulate this model, and see if you can make any predictions about the proportion of white balls in the urn after a large number of draws. Is there a tendency to have a large fraction of balls of the same colour in the long run?

21. It is desired to find the probability that in a bridge deal each player receives an ace. A student argues as follows. It does not matter where the first ace goes. The second ace must go to one of the other three players and this occurs with probability 3/4. Then the next must go to one of two, an event of probability 1/2, and finally the last ace must go to the player who does not have an ace. This occurs with probability 1/4. The probability that all these events occur is the product (3/4)(1/2)(1/4) = 3/32. Is this argument correct?

22. One coin in a collection of 65 has two heads. The rest are fair. If a coin, chosen at random from the lot and then tossed, turns up heads 6 times in a row, what is the probability that it is the two-headed coin?

23. You are given two urns and fifty balls. Half of the balls are white and half are black. You are asked to distribute the balls in the urns with no restriction placed on the number of either type in an urn. How should you distribute the balls in the urns to maximise the probability of obtaining a white ball if an urn is chosen at random and a ball drawn out at random? Justify your answer.

24. A fair coin is thrown n times. Show that the conditional probability of a head on any specified trial, given a total of k heads over the n trials, is k/n ($k > 0$).

25. A coin with probability p for heads is tossed n times. Let E be the event "a head is obtained on the first toss' and F_k the event 'exactly k heads are obtained." For which pairs (n, k) are E and F_k independent?

26. Suppose that A and B are events such that $P(A/B) = P(B/A)$ and $P(A \cup B) = 1$ and $P(A \cap B) > 0$. Prove that $P(A) > 1/2$.

27. In London, half of the days have some rain. The weather forecaster is correct 2/3 of the time, i.e., the probability that it rains, given that she has predicted rain, and the probability that it does not rain, given that she has predicted that it won't rain, are both equal to 2/3. When rain is forecast, Mr. Pickwick takes his umbrella. When rain is not forecast, he takes it with probability 1/3. Find
    (a) the probability that Pickwick has no umbrella, given that it rains.
    (b) the probability that it doesn't rain, given that he brings his umbrella.

28. Probability theory was used in a famous court case: *People v. Collins*. In this case a purse was snatched from an elderly person in a Los Angeles suburb. A couple seen running from the scene were described as a black man with a beard and a moustache and a blond girl with hair in a ponytail. Witnesses

said they drove off in a partly yellow car. Malcolm and Janet Collins were arrested. He was black and though clean shaven when arrested had evidence of recently having had a beard and a moustache. She was blond and usually wore her hair in a ponytail. They drove a partly yellow Lincoln. The prosecution called a professor of mathematics as a witness who suggested that a conservative set of probabilities for the characteristics noted by the witnesses would be as shown in following table.

The prosecution then argued that the probability that all of these characteristics are met by a randomly chosen couple is the product of the probabilities or 1/12,000,000, which is very small. He claimed this was proof beyond a reasonable doubt that the defendants were guilty. The jury agreed and handed down a verdict of guilty of second-degree robbery.

Table: Collins case probabilities.

man with moustache	1/4
girl with blond hair	1/3
girl with ponytail	1/10
black man with beard	1/10
interracial couple in a car	1/1000
partly yellow car	1/10

If you were the lawyer for the Collins couple how would you have countered the above argument?

29. A student is applying to Harvard and Dartmouth. He estimates that he has a probability of .5 of being accepted at Dartmouth and .3 of being accepted at Harvard. He further estimates the probability that he will be accepted by both is .2. What is the probability that he is accepted by Dartmouth if he is accepted by Harvard? Is the event "accepted at Harvard" independent of the event "accepted at Dartmouth"?

30. Luxco, a wholesale lightbulb manufacturer, has two factories. Factory A sells bulbs in lots that consists of 1000 regular and 2000 softglow bulbs each. Random sampling has shown that on the average there tend to be about 2 bad regular bulbs

and 11 bad softglow bulbs per lot. At factory B the lot size is reversed—there are 2000 regular and 1000 softglow per lot—and there tend to be 5 bad regular and 6 bad softglow bulbs per lot.

The manager of factory A asserts, "We're obviously the better producer; our bad bulb rates are .2 per cent and .55 per cent compared to B's .25 per cent and .6 per cent. We're better at both regular and softglow bulbs by half of a tenth of a per cent each."

"Au contraire," counters the manager of B, "each of our 3000 bulb lots contains only 11 bad bulbs, while A's 3000 bulb lots contain 13. So our .37 per cent bad bulb rate beats their .43 per cent."

Who is right?

31. (a) There has been a blizzard and Helen is trying to drive from Woodstock to Tunbridge, which are connected like the top graph in the figure below. Here p and q are the probabilities that the two roads are passable. What is the probability that Helen can get from Woodstock to Tunbridge?

    (b) Now suppose that Woodstock and Tunbridge are connected like the middle graph in the following figure. What now is the probability that she can get from W to T? Note that if we think of the roads as being components of a system, then in (a) and (b) we have computed the reliability of a system whose components are (a) in series and (b) in parallel.

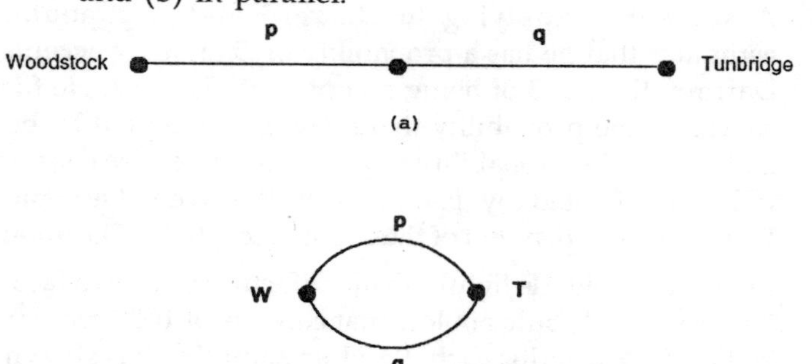

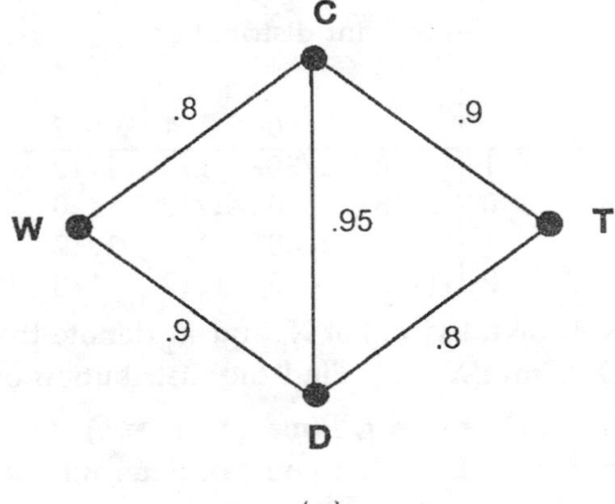

(c)

Figure: From Woodstock to Tunbridge.

(c) Now suppose W and T are connected like the bottom graph in the previous figure. Find the probability of Helen's getting from W to T. *Hint*: If the road from C to D is impassable, it might as well not be there at all; if it is passable, then figure out how to use part (b) twice.

32. Let A_1, A_2, and A_3 be events, and let B_i represent either A_i or its complement $\tilde{A}_i$. Then there are eight possible choices for the triple (B_1, B_2, B_3). Prove that the events A_1, A_2, A_3 are independent if and only if

$$P(B_1 \cap B_2 \cap B_3) = P(B_1)P(B_2)P(B_3),$$

for all eight of the possible choices for the triple (B_1, B_2, B_3).

33. Four women, A, B, C, and D, check their hats, and the hats are returned in a random manner. Let Ω be the set of all possible permutations of A, B, C, D. Let $X_j = 1$ if the jth woman gets her own hat back and 0 otherwise. What is the distribution of X_j? Are the X_i's mutually independent?

34. A box has numbers from 1 to 10. A number is drawn at random. Let X_1 be the number drawn. This number is replaced, and the ten numbers mixed. Asecond number X_2 is drawn. Find the distributions of X_1 and X_2. Are X_1 and X_2 independent? Answer the same questions if the first number is not replaced before the second is drawn.

Table: Joint distribution.

		Y			
		-1	0	1	2
X	-1	0	1/36	1/6	1/12
	0	1/18	0	1/18	0
	1	0	1/36	1/6	1/12
	2	1/12	0	1/12	1/6

35. A die is thrown twice. Let X_1 and X_2 denote the outcomes. Define $X = \min(X_1, X_2)$. Find the distribution of X.

36. Given that $P(X = a) = r$, $P(\max(X, Y) = a) = s$, and $P(\min(X, Y) = a) = t$, show that you can determine $u = P(Y = a)$ in terms of r, s, and t.

37. A fair coin is tossed three times. Let X be the number of heads that turn up on the first two tosses and Y the number of heads that turn up on the third toss. Give the distribution of

    (a) the random variables X and Y.

    (b) the random variable $Z = X + Y$.

    (c) the random variable $W = X - Y$.

38. Assume that the random variables X and Y have the joint distribution given in the previous table.

    (a) What is $P(X \geq 1 \text{ and } Y \leq 0)$?

    (b) What is the conditional probability that $Y \leq 0$ given that $X = 2$?

    (c) Are X and Y independent?

    (d) What is the distribution of $Z = XY$?

39. Two players, A and B, play a series of points in a game with player A winning each point with probability p and player B winning each point with probability $q = 1 - p$. The first player to win N points wins the game. Assume that $N = 3$. Let X be a random variable that has the value 1 if player A wins the series and 0 otherwise. Let Y be a random variable with value the number of points played in a game. Find the distribution of X and Y when $p = 1/2$. Are X and Y

independent in this case? Answer the same questions for the case $p = 2/3$.

40. The letters between Pascal and Fermat, which are often credited with having started probability theory, dealt mostly with the problem of points described in Exercise 39. Pascal and Fermat considered the problem of finding a fair division of stakes if the game must be called off when the first player has won r games and the second player has won s games, with $r < N$ and $s < N$. Let $P(r, s)$ be the probability that player A wins the game if he has already won r points and player B has won s points. Then

    (a) $P(r, N) = 0$ if $r < N$,

    (b) $P(N, s) = 1$ if $s < N$,

    (c) $P(r, s) = pP(r + 1, s) + qP(r, s + 1)$ if $r < N$ and $s < N$,

    and (1), (2), and (3) determine $P(r, s)$ for $r \leq N$ and $s \leq N$. Pascal used these facts to find $P(r, s)$ by working backward: He first obtained $P(N - 1, j)$ for $j = N - 1, N - 2, \ldots, 0$; then, from these values, he obtained $P(N - 2, j)$ for $j = N - 1, N - 2, \ldots, 0$ and, continuing backward, obtained all the values $P(r, s)$. Write a programme to compute $P(r, s)$ for given N, a, b, and p. *Warning*: Follow Pascal and you will be able to run $N = 100$; use recursion and you will not be able to run $N = 20$.

41. Fermat solved the problem of points as follows: He realised that the problem was difficult because the possible ways the play might go are not equally likely. For example, when the first player needs two more games and the second needs three to win, two possible ways the series might go for the first player are WLW and LWLW. These sequences are not equally likely. To avoid this difficulty, Fermat extended the play, adding fictitious plays so that the series went the maximum number of games needed (four in this case). He obtained equally likely outcomes and used, in effect, the Pascal triangle to calculate $P(r, s)$. Show that this leads to a formula for $P(r, s)$ even for the case $p\ 6 \neq 1/2$.

42. The Yankees are playing the Dodgers in a world series. The Yankees win each game with probability .6. What is the probability that the Yankees win the series? (The series is won by the first team to win four games).

43. C. L. Anderson has used Fermat's argument for the problem of points to prove the following result due to J. G. Kingston. You are playing the game of points but, at each point, when you serve you win with probability p, and when your opponent serves you win with probability p. You will serve first, but you can choose one of the following two conventions for serving: for the first convention you alternate service (tennis), and for the second the person serving continues to serve until he loses a point and then the other player serves (racquetball). The first player to win N points wins the game. The problem is to show that the probability of winning the game is the same under either convention.

    (a) Show that, under either convention, you will serve at most N points and your opponent at most $N - 1$ points.

    (b) Extend the number of points to $2N - 1$ so that you serve N points and your opponent serves $N - 1$. For example, you serve any additional points necessary to make N serves and then your opponent serves any additional points necessary to make him serve $N - 1$ points. The winner is now the person, in the extended game, who wins the most points. Show that playing these additional points has not changed the winner.

    (c) Show that (a) and (b) prove that you have the same probability of winning the game under either convention.

44. In the previous problem, assume that $p = 1 - \overline{p}$.

    (a) Show that under either service convention, the first player will win more often than the second player if and only if p > .5.

    (b) In volleyball, a team can only win a point while it is serving. Thus, any individual "play" either ends with a point being awarded to the serving team or with the service changing to the other team. The first team to win N points wins the game. (We ignore here the additional restriction that the winning team must be ahead by at least two points at the end of the game). Assume that each team has the same probability of winning the play when it is serving, i.e., that p = 1 – $\overline{p}$. Show that in this case, the team that serves first will win more than half the time, as long as $p > 0$. (If $p = 0$, then the game never ends). *Hint*: Define p' to be the probability that a team wins the next point, given that it is serving. If we write $q = 1 - p$, then one can show that

$$p' = \frac{p}{1-q^2}.$$

If one now considers this game in a slightly different way, one can see that the second service convention in the preceding problem can be used, with p replaced by p.

45. A poker hand consists of 5 cards dealt from a deck of 52 cards. Let X and Y be, respectively, the number of aces and Kings in a poker hand. Find the joint distribution of X and Y.

46. Let X_1 and X_2 be independent random variables and let $Y_1 = \phi_1(X_1)$ and $Y_2 = \phi_2(X_2)$.

 (a) Show that

$$P(Y_1 = r, Y_2 = s) = \sum_{\substack{\phi_1(a)=r \\ \phi_2(b)=s}} P(X_1 = a, X_2 = b)$$

 (b) Using (a), show that $P(Y_1 = r; Y_2 = s) = P(Y_1 = r)\, P(Y_2 = s)$ so that Y1 and Y2 are independent.

47. Let Ω be the sample space of an experiment. Let E be an event with $P(E) > 0$ and define $m_E(w)$ by $m_E(w) = m(w/E)$. Prove that $m_E(w)$ is a distribution function on E, that is, that $m_E(w) \geq 0$ and that $\sum_{w\in\Omega} m_E(w) = 1$. The function m_E is called the conditional distribution given E.

48. You are given two urns each containing two biased coins. The coins in urn I come up heads with probability p_1, and the coins in urn II come up heads with probability $p_2 \neq p_1$. You are given a choice of (a) choosing an urn at random and tossing the two coins in this urn or (b) choosing one coin from each urn and tossing these two coins. You win a prize if both coins turn up heads. Show that you are better off selecting choice (a).

49. Prove that, if $A_1, A_2, \ldots, A_n$ are independent events defined on a sample space Ω and if $0 < P(A_j) < 1$ for all j, then Ω must have at least 2^n points.

50. Prove that if

$$P(A/C) \geq P(B/C) \text{ and } P(A/\tilde{C}) \geq P(B/\tilde{C}),$$

then $P(A) \geq P(B)$.

51. A coin is in one of n boxes. The probability that it is in the ith box is p_i. If you search in the ith box and it is there, you find it with probability a_i. Show that the probability p that the coin is in the jth box, given that you have looked in the ith box and not found it, is

$$p = \begin{cases} p_j/(1-a_i p_i), & \text{if } j \neq i, \\ (1-a_i)p_i/(1-a_i p_i), & \text{if } j = i. \end{cases}$$

52. George Wolford has suggested the following variation on the Linda problem. The registrar is carrying John and Mary's registration cards and drops them in a puddle. When he pickes them up he cannot read the names but on the first card he picked up he can make out Mathematics 23 and Government 35, and on the second card he can make out only Mathematics 23. He asks you if you can help him decide which card belongs to Mary. You know that Mary likes government but does not like mathematics. You know nothing about John and assume that he is just a typical Dartmouth student. From this you estimate:

P (Mary takes Government 35) = .5,
P (Mary takes Mathematics 23) = .1,

P (John takes Government 35) = .3,
P (John takes Mathematics 23) = .2.

Assume that their choices for courses are independent events. Show that the card with Mathematics 23 and Government 35 showing is more likely to be Mary's than John's. The conjunction fallacy referred to in the Linda problem would be to assume that the event "Mary takes Mathematics 23 and Government 35" is more likely than the event "Mary takes Mathematics 23." Why are we not making this fallacy here?

53. A deck of playing cards can be described as a Cartesian product

$$\text{Deck} = \text{Suit} \times \text{Rank},$$

where Suit = $\{\clubsuit, \diamondsuit, \heartsuit, \spadesuit\}$ and Rank = $\{2, 3, \ldots, 10, J, Q, K, A\}$. This just means that every card may be thought of as an ordered pair like $(\diamondsuit, 2)$. By a suit event we mean any event A contained in Deck which is described in terms of Suit alone. For instance, if A is "the suit is red," then

$$A = \{\diamondsuit, \heartsuit\} \times \text{Rank},$$

so that A consists of all cards of the form $(\diamondsuit, r)$ or $(\heartsuit, r)$ where r is any rank. Similarly, a *rank event* is any event described in terms of rank alone.

(a) Show that if A is any suit event and B any rank event, then A and B are independent. (We can express this briefly by saying that suit and rank are independent).

(b) Throw away the ace of spades. Show that now no non-trivial (i.e., neither empty nor the whole space) suit event A is independent of any non-trivial rank event B. *Hint*: Here independence comes down to

$$c/51 = (a/51) \cdot (b/51),$$

where a, b, c are the respective sizes of A, B and $A \cap B$. It follows that 51 must divide ab, hence that 3 must divide one of a and b, and 17 the other. But the possible sizes for suit and rank events preclude this.

(c) Show that the deck in (b) nevertheless does have pairs A, B of non-trivial independent events. *Hint*: Find 2 events A and B of sizes 3 and 17, respectively, which intersect in a single point.

(d) Add a joker to a full deck. Show that now there is no pair A, B of non-trivial independent events. *Hint*: See the hint in (b); 53 is prime.

The following problems are suggested by Stanley Gudder in his article "Do Good Hands Attract?" He says that event A attracts event B if $P(B/A) > P(B)$ and repels B if $P(B/A) < P(B)$.

54. Let R_i be the event that the ith player in a poker game has a royal flush. Show that a royal flush (A, K, Q, J, 10 of one suit) attracts another royal flush, that is $P(R_2/R_1) > P(R_2)$. Show that a royal flush repels full houses.

55. Prove that A attracts B if and only if B attracts A. Hence we can say that A and B are mutually attractive if A attracts B.

56. Prove that A neither attracts nor repels B if and only if A and B are independent.

57. Prove that A and B are mutually attractive if and only if $P(B/A) > P(B/\tilde{A})$.

58. Prove that if A attracts B, then A repels $\tilde{B}$.

59. Prove that if A attracts both B and C, and A repels $B \cap C$, then A attracts $B \cup C$. Is there any example in which A attracts both B and C and repels $B \cup C$?

60. Prove that if $B_1, B, \ldots, B_n$ are mutually disjoint and collectively exhaustive, and if A attracts some B_i, then A must repel some B_j.

61. (a)Suppose that you are looking in your desk for a letter from some time ago. Your desk has eight drawers, and you assess the probability that it is in any particular drawer is 10 per cent (so there is a 20 per cent chance that it is not in the desk at all). Suppose now that you start searching systematically through your desk, one drawer at a time. In addition, suppose

that you have not found the letter in the first i drawers, where $0 \le i \le 7$. Let p_i denote the probability that the letter will be found in the next drawer, and let q_i denote the probability that the letter will be found in some subsequent drawer (both p_i and q_i are conditional probabilities, since they are based upon the assumption that the letter is not in the first i drawers). Show that the p_i's increase and the q_i's decrease. (This problem is from Falk *et al*).

(b) The following data appeared in an article in the Wall Street Journal. For the ages 20, 30, 40, 50, and 60, the probability of a woman in the US developing cancer in the next ten years is 0.5 per cent, 1.2 per cent, 3.2 per cent, 6.4 per cent, and 10.8 per cent, respectively. At the same set of ages, the probability of a woman in the US eventually developing cancer is 39.6 per cent, 39.5 per cent, 39.1 per cent, 37.5 per cent, and 34.2 per cent, respectively. Do you think that the problem in part (a) gives an explanation for these data?

62. Here are two variations of the Monty Hall problem that are discussed by Granberg.

(a) Suppose that everything is the same except that Monty forgot to find out in advance which door has the car behind it. In the spirit of "the show must go on," he makes a guess at which of the two doors to open and gets lucky, opening a door behind which stands a goat. Now should the contestant switch?

(b) You have observed the show for a long time and found that the car is put behind door A 45 per cent of the time, behind door B 40 per cent of the time and behind door C 15 per cent of the time. Assume that everything else about the show is the same. Again you pick door A. Monty opens a door with a goat and offers to let you switch. Should you? Suppose you knew in advance that Monty was going to give you a chance to switch. Should you have initially chosen door A?

Continuous Conditional Probability

Thus, for example, if X is a continuous random variable with density function $f(x)$, and if E is an event with positive probability, we define a conditional density function by the formula

$$f(x/E) = \begin{cases} f(x)/P(E), & \text{if } x \in E, \\ 0, & \text{if } x \notin E. \end{cases}$$

Then for any event F, we have

$$P(F/E) = \int_F f(x/E)dx.$$

The expression $P(F/E)$ is called the conditional probability of F given E. It is easy to obtain an alternative expression for this probability:

$$P(F/E) = \int_F f(x/E)dx = \int_{E\cap F} \frac{f(x)}{P(E)}dx = \frac{P(E\cap F)}{P(E)}.$$

We can think of the conditional density function as being 0 except on E, and normalised to have integral 1 over E. Note that if the original density is a uniform density corresponding to an experiment in which all events of equal size are equally likely, then the same will be true for the conditional density.

Example: In the spinner experiment, suppose we know that the spinner has stopped with head in the upper half of the circle, $0 \le x \le 1/2$. What is the probability that $1/6 \le x \le 1/3$?

Here $E = [0, 1/2]$, $F = [1/6, 1/3]$, and $F \cap E = F$. Hence

$$\begin{aligned} P(F/E) &= \frac{P(F\cap E)}{P(E)} \\ &= \frac{1/6}{1/2} \\ &= \frac{1}{3}, \end{aligned}$$

which is reasonable, since F is $1/3$ the size of E. The conditional density function here is given by

$$f(x/E) = \begin{cases} 2, & \text{if } 0 \le x < 1/2, \\ 0, & \text{if } 1/2 \le x < 1. \end{cases}$$

Thus the conditional density function is non-zero only on [0, 1/2], and is uniform there.

Example: In the dart game, suppose we know that the dart lands in the upper half of the target. What is the probability that its distance from the centre is less than 1/2?

Here $E = \{(x, y) : y \ge 0\}$, and $F = f(x, y) : x^2 + y^2 < (1/2)^2\}$. Hence,

$$P(F/E) = \frac{P(F \cap E)}{P(E)} = \frac{(1/\pi)[(1/2)(\pi/4)]}{(1/\pi)(\pi/2)}$$

$$= 1/4.$$

Here again, the size of $F \cap E$ is 1/4 the size of E. The conditional density function

$$f((x, y)/E) = \begin{cases} f(x,y)/P(E) = 2/\pi, & \text{if } (x,y) \in E, \\ 0, & \text{if } (x,y) \notin E. \end{cases}$$

Example: We return to the exponential density. We suppose that we are observing a lump of plutonium-239. Our experiment consists of waiting for an emission, then starting a clock, and recording the length of time X that passes until the next emission. Experience has shown that X has an exponential density with some parameter λ, which depends upon the size of the lump. Suppose that when we perform this experiment, we notice that the clock reads r seconds, and is still running. What is the probability that there is no emission in a further s seconds?

Let $G(t)$ be the probability that the next particle is emitted after time t. Then

$$G(t) = \int_t^\infty \lambda e^{-\lambda x} dx$$

$$= -e^{-\lambda x} \Big|_t^\infty = e^{-\lambda t}.$$

Let E be the event "the next particle is emitted after time r" and F the event "the next particle is emitted after time $r + s$." Then

$$\begin{aligned} P(F/E) &= \frac{P(F \cap E)}{P(E)} \\ &= \frac{G(r+s)}{G(r)} \\ &= \frac{e^{-\lambda(r+s)}}{e^{-\lambda r}} \\ &= e^{-\lambda s}. \end{aligned}$$

This tells us the rather surprising fact that the probability that we have to wait s seconds more for an emission, given that there has been no emission in r seconds, is independent of the time r. This property is called the memoryless property. When trying to model various phenomena, this property is helpful in deciding whether the exponential density is appropriate.

The fact that the exponential density is memoryless means that it is reasonable to assume if one comes upon a lump of a radioactive isotope at some random time, then the amount of time until the next emission has an exponential density with the same parameter as the time between emissions. A well-known example, known as the "bus paradox," replaces the emissions by buses. The apparent paradox arises from the following two facts: 1) If you know that, on the average, the buses come by every 30 minutes, then if you come to the bus stop at a random time, you should only have to wait, on the average, for 15 minutes for a bus, and 2) Since the buses arrival times are being modelled by the exponential density, then no matter when you arrive, you will have to wait, on the average, for 30 minutes for a bus.

If one makes a reasonable assumption on the distribution of the interarrival times, then the average waiting time is more nearly one-half the average interarrival time.

Independent Events

If E and F are two events with positive probability in a continuous sample space, then, as in the case of discrete sample spaces, we define E and F to be independent if $P(E/F) = P(E)$ and $P(F/E) = P(F)$. As before, each of the above equations imply the other, so that to see whether two events are independent, only one of these equations must be checked. It is also the case that, if E and F are independent, then $P(E \cap F) = P(E)P(F)$.

Example: In the dart game, let E be the event that the dart lands in the upper half of the target ($y \geq 0$) and F the event that the dart lands in the right half of the target ($x \geq 0$). Then $P(E \cap F)$ is the probability that the dart lies in the first quadrant of the target, and

$$\begin{aligned} P(E \cap F) &= \frac{1}{\pi}\int_{E \cap F} 1dxdy \\ &= \text{Area } (E \cap F) \\ &= \text{Area } (E)\text{Area } (F) \\ &= \left(\frac{1}{\pi}\int_{E} 1dxdy\right)\left(\frac{1}{\pi}\int_{F} 1dxdy\right) \\ &= P(E)P(F) \end{aligned}$$

so that E and F are independent. What makes this work is that the events E and F are described by restricting different coordinates. This idea is made more precise below.

Joint Density and Cumulative Distribution Functions

In a manner analogous with discrete random variables, we can define joint density functions and cumulative distribution functions for multidimensional continuous random variables.

Definition: Let $X_1, X_2, \ldots, X_n$ be continuous random variables associated with an experiment, and let $\overline{X} = (X_1, X_2, \ldots, X_n)$. Then the joint cumulative distribution function of $\overline{X}$ is defined by

$$F(x_1, x_2, \ldots, x_n) = P(X_1 \leq x_1, X_2 \leq x_2, \ldots, X_n \leq x_n).$$

The joint density function of $\overline{X}$ satisfies the following equation:

$$F(x_1, x_2, \ldots, x_n) = \int_{-\infty}^{x_1}\int_{-\infty}^{x_2}\cdots\int_{-\infty}^{x_n} f(t_1, t_2, \ldots t_n)\, dt_n dt_{n-1} \cdots dt_1.$$

It is straightforward to show that, in the above notation,

$$F(x_1, x_2, \ldots, x_n) = \frac{\partial^n F(x_1, x_2 \cdots, x_n)}{\partial x_1 \partial x_2 \cdots \partial x_n}. \tag{4.4}$$

Independent Random Variables

As with discrete random variables, we can define mutual independence of continuous random variables.

Definition: Let $X_1, X_2, \ldots, X_n$ be continuous random variables with cumulative distribution functions $F_1(x), F_2(x), \ldots, F_n(x)$. Then these random variables are mutually independent if

$$F(x_1, x_2, \ldots, x_n) = F_1(x_1)F_2(x_2)\ldots F_n(x_n)$$

for any choice of $x_1, x_2, \ldots, x_n$. Thus, if $X_1, X_2, \ldots, X_n$ are mutually independent, then the joint cumulative distribution function of the random variable $X = (X_1, X_2, \ldots, X_n)$ is just the product of the individual cumulative distribution functions. When two random variables are mutually independent, we shall say more briefly that they are independent. Using equation 4.4, the following theorem can easily be shown to hold for mutually independent continuous random variables.

Theorem: Let $X_1, X_2, \ldots, X_n$ be continuous random variables with density functions $f_1(x), f_2(x), \ldots, f_n(x)$. Then these random variables are *mutually independent* if and only if

$$f(x_1, x_2, \ldots, x_n) = f_1(x_1)f_2(x_2)\ldots f_n(x_n)$$

for any choice of $x_1, x_2, \ldots, x_n$.

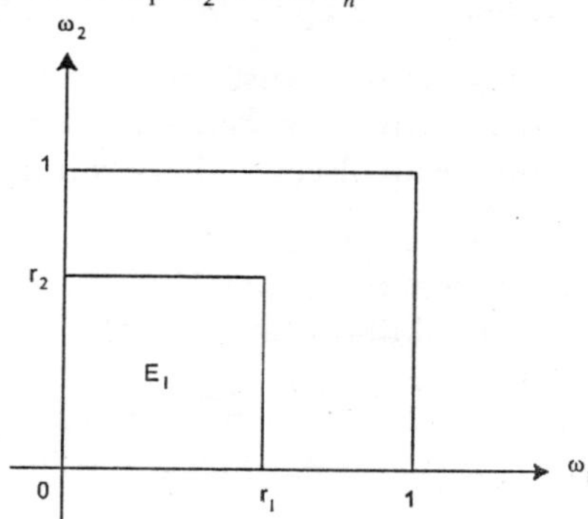

Figure: X_1 and X_2 are independent.

Let's look at some examples.

Example: In this example, we define three random variables, X_1, X_2, and X_3. We will show that X_1 and X_2 are independent, and that X_1 and X_3 are not independent. Choose a point $w = (w_1, w_2)$

at random from the unit square. Set $X_1 = w_1^2$, $X_2 = w_2^2$, and $X_3 = w_1 + w_2$. Find the joint distributions $F_{12}(r_1, r_2)$ and $F_{23}(r_2, r_3)$.

We have already seen that

$$\begin{aligned} F_1(r_1) &= P(-\infty < X_1 \le r_1) \\ &= \sqrt{r_1}, \qquad \text{if } 0 \le r_1 \le 1, \end{aligned}$$

and similarly,

$$F_2(r_2) = \sqrt{r_2},$$

if $0 \le r_2 \le 1$. Now we have

$$\begin{aligned} F_{12}(r_1, r_2) &= P(X_1 \le r_1 \text{ and } X_2 \le r_2) \\ &= P(w_1 \le \sqrt{r_1} \text{ and } w_2 \le \sqrt{r_2}) \\ &= \text{Area}(E_1) \\ &= \sqrt{r_1}\sqrt{r_2} \\ &= F_1(r_1)F_2(r_2). \end{aligned}$$

In this case $F_{12}(r_1, r_2) = F_1(r_1)F_2(r_2)$ so that X_1 and X_2 are independent. On the other hand, if $r_1 = 1/4$ and $r_3 = 1$, then

$$F_{13}(1/4, 1) = P(X_1 \le 1/4, X_3 \le 1).$$

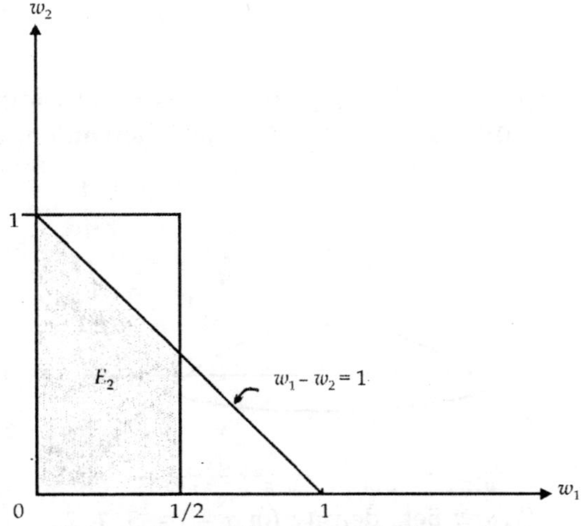

w_2

1

E_2

$w_1 - w_2 = 1$

w_1

0 1/2 1

Figure: X_1 and X_3 are not independent.

$$= P(w_1 \le 1/2, w_1 + w_2 \le 1)$$

$$= \text{Area } (E_2)$$

$$= \frac{1}{2} - \frac{1}{8} = \frac{3}{8}.$$

Now recalling that

$$F_3(r_3) = \begin{cases} 0, & \text{if } r_3 < 0, \\ (1/2)r_3^2, & \text{if } 0 \le r_3 \le 1, \\ 1-(1/2)(2-r_3)^2, & \text{if } 1 \le r_3 \le 2, \\ 1, & \text{if } 2 < r_3, \end{cases}$$

we have $F_1(1/4)F_3(1) = (1/2)(1/2) = 1/4$. Hence, X_1 and X_3 are not independent random variables. A similar calculation shows that X_2 and X_3 are not independent either.

Although we shall not prove it here, the following theorem is a useful one. The statement also holds for mutually independent discrete random variables. A proof may be found in Renyi.

Theorem: Let $X_1, X_2, \ldots, X_n$ be mutually independent continuous random variables and let $\phi_1(x), \phi_2(x), \ldots, \phi_n(x)$ be continuous functions. Then $\phi_1(X_1), \phi_2(X_2), \ldots, \phi_n(X_n)$ are mutually independent.

Independent Trials

Using the notion of independence, we can now formulate for continuous sample spaces the notion of independent trials.

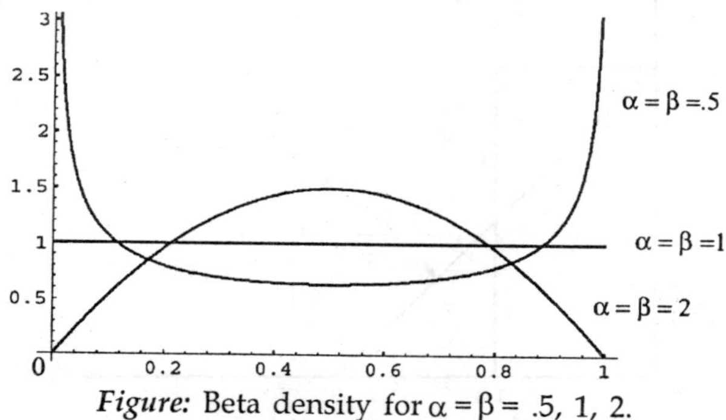

Figure: Beta density for α = β = .5, 1, 2.

Definition: A sequence $X_1, X_2, \ldots, X_n$ of random variables X_i that are mutually independent and have the same density is called an independent trials process.

As in the case of discrete random variables, these independent trials processes arise naturally in situations where an experiment described by a single random variable is repeated n times.

Beta Density

We consider next an example which involves a sample space with both discrete and continuous coordinates. For this example, we shall need a new density function called the beta density. This density has two parameters α, β and is defined by

$$B(\alpha,\beta,x) = \begin{cases} (1/B(\alpha,\beta))x^{\alpha-1}(1-x)^{\beta-1}, & \text{if } 0 \le x \le 1, \\ 0, & \text{otherwise}. \end{cases}$$

Here α and β are any positive numbers, and the beta function $B(\alpha,\beta)$ is given by the area under the graph of $x^{\alpha-1}(1-x)^{\beta-1}$ between 0 and 1:

$$B(\alpha,\beta) = \int_0^1 x^{\alpha-1}(1-x)^{\beta-1}\,dx.$$

Note that when $\alpha = \beta = 1$ the beta density if the uniform density. When α and β are greater than 1 the density is bell-shaped, but when they are less than 1 it is *U*-shaped as suggested by the examples in the figure above.

We shall need the values of the beta function only for integer values of α and β, and in this case

$$B(\alpha,\beta) = \frac{(\alpha-1)!(\beta-1)!}{(\alpha+\beta-1)!}.$$

Example: In medical problems it is often assumed that a drug is effective with a probability x each time it is used and the various trials are independent, so that one is, in effect, tossing a biased coin with probability x for heads. Before further experimentation, you do not know the value x but past experience might give some information about its possible values. It is natural to represent this information by sketching a density function to determine a distribution for x. Thus, we are considering x to be a continuous random variable, which takes on values between 0 and 1. If you have no knowledge at all, you would sketch the uniform density. If past experience suggests that x is very likely to be near 2/3 you would sketch a density with maximum at 2/3 and a spread reflecting your uncertainly in the estimate of 2/3. You would then want to find a density function that reasonably fits your sketch. The beta densities provide a class of densities that can be fit to most sketches you might make. For example, for $\alpha > 1$ and $\beta > 1$ it is bell-shaped with the parameters α and β determining its peak and its spread.

Assume that the experimenter has chosen a beta density to describe the state of his knowledge about x before the experiment. Then he gives the drug to n subjects and records the number i of successes. The number i is a discrete random variable, so we may conveniently describe the set of possible outcomes of this experiment by referring to the ordered pair (x, i).

We let $m(i \mid x)$ denote the probability that we observe i successes given the value of x. By our assumptions, $m(i \mid x)$ is the binomial distribution with probability x for success:

$$m(i \mid x) = b(n,x,i) = \binom{n}{i} x^i (1-x)^j ,$$

where $j = n - i$.

If x is chosen at random from [0, 1] with a beta density $B(\alpha, \beta, x)$, then the density function for the outcome of the pair (x, i) is

$$\begin{aligned} f(x,i) &= m(i \mid x) B(\alpha,\beta,x) \\ &= \binom{n}{i} x^i (1-x)^j \frac{1}{B(\alpha,\beta)} x^{\alpha-1} (1-x)^{\beta-1} \\ &= \binom{n}{i} \frac{1}{B(\alpha,\beta)} x^{\alpha+i-1} (1-x)^{\beta+j-1} . \end{aligned}$$

Now let $m(i)$ be the probability that we observe i successes *not* knowing the value of x. Then

$$\begin{aligned} m(i) &= \int_0^1 m(i \mid x) B(\alpha,\beta,x)\,dx \\ &= \binom{n}{i} \frac{1}{B(\alpha,\beta)} \int_0^1 x^{\alpha+i-1}(1-x)^{\beta+j-1}\,dx \\ &= \binom{n}{i} \frac{B(\alpha+i,\beta+j)}{B(\alpha,\beta)}. \end{aligned}$$

Hence, the probability density $f(x \mid i)$ for x, given that i successes were observed, is

$$\begin{aligned} f(x \mid i) &= \frac{f(x,i)}{m(i)} \\ &= \frac{x^{\alpha+i-1}(1-x)^{\beta+j-1}}{B(\alpha+i,\beta+j)}, \end{aligned} \tag{4.5}$$

that is, $f(x \mid i)$ is another beta density. This says that if we observe i successes and j failures in n subjects, then the new density for the probability that the drug is effective is again a beta density but with parameters $\alpha + i$, $\beta + j$.

Now we assume that before the experiment we choose a beta density with parameters α and β, and that in the experiment we obtain i successes in n trials. We have just seen that in this case, the new density for x is a beta density with parameters $\alpha + i$ and $\beta + j$.

Now we wish to calculate the probability that the drug is effective on the next subject. For any particular real number t between 0 and 1, the probability that x has the value t is given by the expression in equation 4.5. Given that x has the value t, the probability that the drug is effective on the next subject is just t. Thus, to obtain the probability that the drug is effective on the next subject, we integrate the product of the expression in equation 4.5 and t overall possible values of t. We obtain:

$$\begin{aligned} &\frac{1}{B(\alpha+i,\beta+j)} \int_0^1 t \cdot t^{\alpha+i-1}(1-t)^{\beta+j-1}\,dt \\ &= \frac{B(\alpha+i+1,\beta+j)}{B(\alpha+i,\beta+j)} \end{aligned}$$

$$= \frac{(\alpha+i)!(\beta+j-1)!}{(\alpha+\beta+i+j)!} \cdot \frac{(\alpha+\beta+i+j-1)!}{(\alpha+i-1)!(\beta+j-1)!}$$

$$= \frac{\alpha+i}{\alpha+\beta+n}.$$

If n is large, then our estimate for the probability of success after the experiment is approximately the proportion of successes observed in the experiment, which is certainly a reasonable conclusion.

The next example is another in which the true probabilities are unknown and must be estimated based upon experimental data.

Example: You are in a casino and confronted by two slot machines. Each machine pays-off either 1 dollar or nothing. The probability that the first machine pays-off a dollar is x and that the second machine pays-off a dollar is y. We assume that x and y are random numbers chosen independently from the interval [0, 1] and unknown to you. You are permitted to make a series of ten plays, each time choosing one machine or the other. How should you choose to maximise the number of times that you win?

One strategy that sounds reasonable is to calculate, at every stage, the probability that each machine will pay off and choose the machine with the higher probability. Let win(i), for $i = 1$ or 2, be the number of times that you have won on the ith machine. Similarly, let lose(i) be the number of times you have lost on the ith machine. Then, from Example 4.23, the probability p(i) that you win if you choose the ith machine is

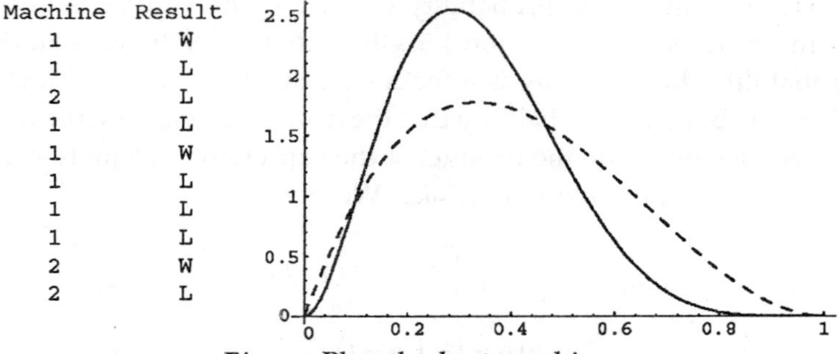

Figure: Play the best machine.

$$p(i) = \frac{\text{win}(i)+1}{\text{win}(i)+\text{lose}(i)+2}.$$

Thus, if $p(1) > p(2)$ you would play machine 1 and otherwise you would play machine 2. We have written a programme *TwoArm* to simulate this experiment. In the programme, the user specifies the initial values for x and y (but these are unknown to the experimenter). The programme calculates at each stage the two conditional densities for x and y, given the outcomes of the previous trials, and then computes $p(i)$, for $i = 1, 2$. It then chooses the machine with the highest value for the probability of winning for the next play. The programme prints the machine chosen on each play and the outcome of this play. It also plots the new densities for x (solid line) and y (dotted line), showing only the current densities. We have run the programme for ten plays for the case $x = .6$ and $y = .7$. The result is shown in the previous figure.

The run of the programme shows the weakness of this strategy. Our initial probability for winning on the better of the two machines is .7. We start with the poorer machine and our outcomes are such that we always have a probability greater than .6 of winning and so we just keep playing this machine even though the other machine is better. If we had lost on the first play we would have switched machines.

Our final density for y is the same as our initial density, namely, the uniform density. Our final density for x is different and reflects a much more accurate knowledge about x. The computer did pretty well with this strategy, winning seven out of the ten trials, but ten trials are not enough to judge whether this is a good strategy in the long run.

Another popular strategy is the play-the-winner strategy. As the name suggests, for this strategy we choose the same machine when we win and switch machines when we lose. The programme TwoArm will simulate this strategy as well. In the following figure, we show the results of running this programme with the play-the-winner strategy and the same true probabilities of .6 and .7 for the two machines. After ten plays our densities for the unknown probabilities of winning suggest to us that the second machine is indeed the better of the two. We again won seven out of the ten trials.

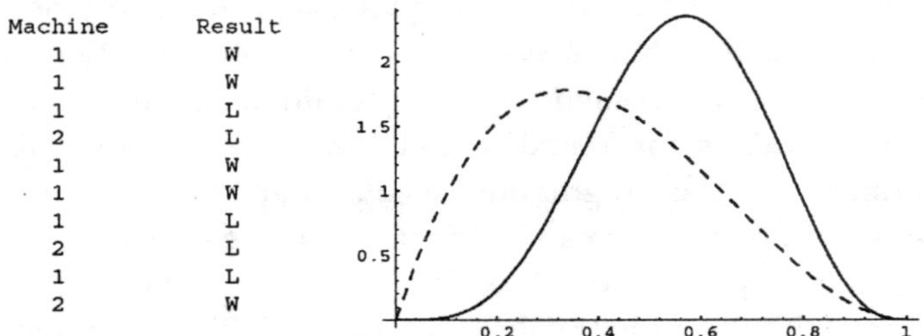

Figure: Play the winner.

Neither of the strategies that we simulated is the best one in terms of maximising our average winnings. This best strategy is very complicated but is reasonably approximated by the play-the-winner strategy. Variations on this example have played an important role in the problem of clinical tests of drugs where experimenters face a similar situation.

Exercises

1. Pick a point x at random (with uniform density) in the interval $[0, 1]$. Find the probability that $x > 1/2$, given that
   (a) $x > 1/4$.
   (b) $x < 3/4$.
   (c) $|x - 1/2| < 1/4$.
   (d) $x^2 - x + 2/9 < 0$.
2. A radioactive material emits α-particles at a rate described by the density function
$$f(t) = .1e^{-.1t}.$$
   Find the probability that a particle is emitted in the first 10 seconds, given that
   (a) no particle is emitted in the first second.
   (b) no particle is emitted in the first 5 seconds.
   (c) a particle is emitted in the first 3 seconds.
   (d) a particle is emitted in the first 20 seconds.

3. The Acme Super light bulb is known to have a useful life described by the density function

$$f(t) = .01e^{-.01t},$$

where time t is measured in hours.

(a) Find the failure rate of this bulb.

(b) Find the reliability of this bulb after 20 hours.

(c) Given that it lasts 20 hours, find the probability that the bulb lasts another 20 hours.

(d) Find the probability that the bulb burns out in the forty-first hour, given that it lasts 40 hours.

4. Suppose you toss a dart at a circular target of radius 10 inches. Given that the dart lands in the upper half of the target, find the probability that

(a) it lands in the right half of the target.

(b) its distance from the centre is less than 5 inches.

(c) its distance from the centre is greater than 5 inches.

(d) it lands within 5 inches of the point (0, 5).

5. Suppose you choose two numbers x and y, independently at random from the interval [0, 1]. Given that their sum lies in the interval [0, 1], find the probability that

(a) $|x - y| < 1$.

(b) $xy < 1/2$.

(c) $\max\{x, y\} < 1/2$.

(d) $x^2 + y^2 < 1/4$.

(e) $x > y$.

6. Find the conditional density functions for the following experiments.

(a) A number x is chosen at random in the interval [0, 1], given that $x > 1/4$.

(b) A number t is chosen at random in the interval $[0, \infty]$ with exponential density e^{-t}, given that $1 < t < 10$.

(c) A dart is thrown at a circular target of radius 10 inches, given that it falls in the upper half of the target.

(d) Two numbers x and y are chosen at random in the interval [0, 1], given that $x > y$.

7. Let x and y be chosen at random from the interval [0, 1]. Show that the events $x > 1/3$ and $y > 2/3$ are independent events.

8. Let x and y be chosen at random from the interval [0, 1]. Which pairs of the following events are independent?

(a) $x > 1/3$.

(b) $y > 2/3$.

(c) $x > y$.

(d) $x + y < 1$.

9. Suppose that X and Y are continuous random variables with density functions $f_X(x)$ and $f_Y(y)$, respectively. Let $f(x, y)$ denote the joint density function of (X, Y). Show that

$$\int_{-\infty}^{\infty} f(x,y)\,dy = f_X(x),$$

and

$$\int_{-\infty}^{\infty} f(x,y)\,dx = f_Y(y),$$

10. If you take a stick of unit length and break it into three pieces, choosing the breaks at random (i.e., choosing two real numbers independently and uniformly from [0, 1]), then the probability that the three pieces form a triangle is 1/4. Consider now a similar experiment: First break the stick at random, then break the longer piece at random. Show that the two experiments are actually quite different, as follows:

(a) Write a programme which simulates both cases for a run of 1000 trials, prints out the proportion of successes for each run, and repeats this process ten times. (Call a trial a success if the three pieces do form a triangle.) Have your programme pick (x, y) at random in the unit square, and in each case use x and y to find the two breaks. For each experiment, have it plot (x, y) if (x, y) gives a success.

(b) Show that in the second experiment the theoretical probability of success is actually 2 log 2 – 1.

11. A coin has an unknown bias p that is assumed to be uniformly distributed between 0 and 1. The coin is tossed n times and heads turns up j times and tails turns up k times. We have seen that the probability that heads turns up next time is

$$\frac{j+1}{n+2}.$$

Show that this is the same as the probability that the next ball is black for the Polya urn model. Use this result to explain why, in the Polya urn model, the proportion of black balls does not tend to 0 or 1 as one might expect but rather to a uniform distribution on the interval [0, 1].

12. Previous experience with a drug suggests that the probability p that the drug is effective is a random quantity having a beta density with parameters $\alpha = 2$ and $\beta = 3$. The drug is used on ten subjects and found to be successful in four out of the ten patients. What density should we now assign to the probability p? What is the probability that the drug will be successful the next time it is used?

13. Write a programme to allow you to compare the strategies play-the-winner and play-the-best-machine for the two-armed bandit problem. Have your programme determine the initial payoff probabilities for each machine by choosing a pair of random numbers between 0 and 1. Have your programme carry out 20 plays and keep track of the number of wins for each of the two strategies. Finally, have your programme make 1000 repetitions of the 20 plays and compute the average winning per 20 plays. Which strategy seems to be the best? Repeat these simulations with 20 replaced by 100. Does your answer to the above question change?

14. Consider the two-armed bandit problem. Bruce Barnes proposed the following strategy, which is a variation on the

play-the-best-machine strategy. The machine with the greatest probability of winning is played unless the following two conditions hold: a) the difference in the probabilities for winning is less than .08, and b) the ratio of the number of times played on the more often played machine to the number of times played on the less often played machine is greater than 1.4. If the above two conditions hold, then the machine with the smaller probability of winning is played. Write a programme to simulate this strategy. Have your programme choose the initial payoff probabilities at random from the unit interval [0, 1], make 20 plays, and keep track of the number of wins. Repeat this experiment 1000 times and obtain the average number of wins per 20 plays. Implement a second strategy—for example, play-the-best-machine or one of your own choice, and see how this second strategy compares with Bruce's on average wins.

Paradoxes

One must be very careful in dealing with problems involving conditional probability. The reader will recall that in the Monty Hall problem, if the contestant chooses the door with the car behind it, then Monty has a choice of doors to open. We made an assumption that in this case, he will choose each door with probability 1/2. We then noted that if this assumption is changed, the answer to the original question changes. In this section, we will study other examples of the same phenomenon.

Example: Consider a family with two children. Given that one of the children is a boy, what is the probability that both children are boys?

One way to approach this problem is to say that the other child is equally likely to be a boy or a girl, so the probability that both children are boys is 1/2. The "textbook" solution would be to draw the tree diagram and then form the conditional tree by deleting paths to leave only those paths that are consistent with the given information.

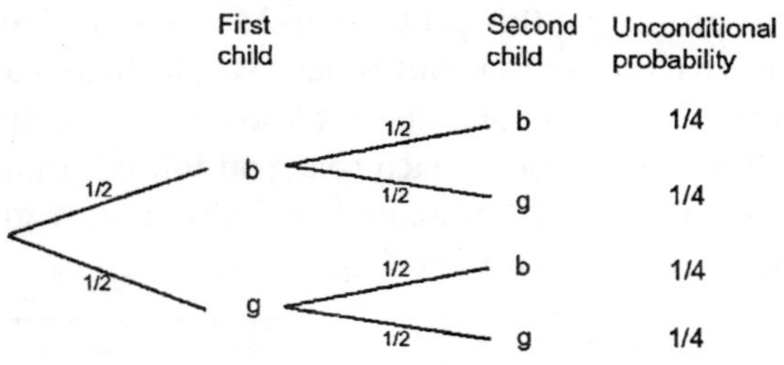

First child		Second child	Unconditional probability	Conditional probability
1/2 b	1/2	b	1/4	1/3
	1/2	g	1/4	1/3
1/2 g	1/2	b	1/4	1/3

Figure: Family tree.

The result is shown in the figure above. We see that the probability of two boys given a boy in the family is not 1/2 but rather 1/3.

This problem and others like it are discussed in Bar-Hillel and Falk. These authors stress that the answer to conditional probabilities of this kind can change depending upon how the information given was actually obtained. For example, they show that 1/2 is the correct answer for the following scenario.

Example: Mr. Smith is the father of two. We meet him walking along the street with a young boy whom he proudly introduces as his son. What is the probability that Mr. Smith's other child is also a boy?

As usual we have to make some additional assumptions. For example, we will assume that if Mr. Smith has a boy and a girl, he is equally likely to choose either one to accompany him on his walk. In the following figure, we show the tree analysis of this problem and we see that 1/2 is, indeed, the correct answer.

Example: It is not so easy to think of reasonable scenarios that would lead to the classical 1/3 answer. An attempt was made by

Stephen Geller in proposing this problem to Marilyn vos Savant. Geller's problem is as follows: A shopkeeper says she has two new baby beagles to show you, but she doesn't know whether they're both male, both female, or one of each sex. You tell her that you want only a male, and she telephones the fellow who's giving them a bath. "Is at least one a male?" she asks.

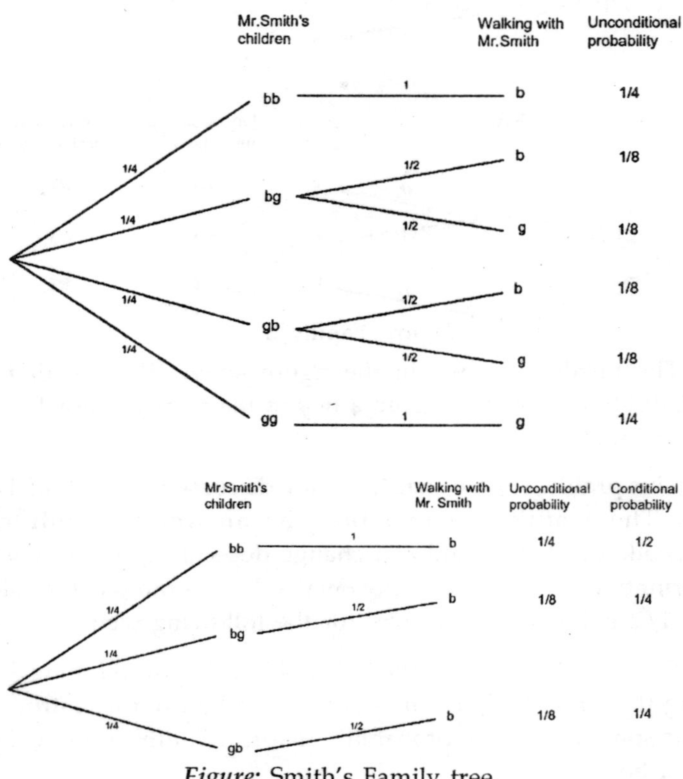

Figure: Smith's Family tree.

"Yes," she informs you with a smile. What is the probability that the other one is male?

The reader is asked to decide whether the model which gives an answer of 1/3 is a reasonable one to use in this case.

In the preceding examples, the apparent paradoxes could easily be resolved by clearly stating the model that is being used and the assumptions that are being made. We now turn to some examples in which the paradoxes are not so easily resolved.

Example: Two envelopes each contain a certain amount of money. One envelope is given to Ali and the other to Baba and they are told that one envelope contains twice as much money as the other. However, neither knows who has the larger prize. Before anyone has opened their envelope, Ali is asked if she would like to trade her envelope with Baba. She reasons as follows: Assume that the amount in my envelope is x. If I switch, I will end up with $x=2$ with probability 1/2, and $2x$ with probability 1/2. If I were given the opportunity to play this game many times, and if I were to switch each time, I would, on average, get

$$\frac{1}{2}\frac{x}{2}+\frac{1}{2}2x = \frac{5}{4}x.$$

This is greater than my average winnings if I didn't switch.

Of course, Baba is presented with the same opportunity and reasons in the same way to conclude that he too would like to switch. So they switch and each thinks that his/her net worth just went up by 25 per cent.

Since neither has yet opened any envelope, this process can be repeated and so again they switch. Now they are back with their original envelopes and yet they think that their fortune has increased 25 per cent twice. By this reasoning, they could convince themselves that by repeatedly switching the envelopes, they could become arbitrarily wealthy. Clearly, something is wrong with the above reasoning, but where is the mistake?

One of the tricks of making paradoxes is to make them slightly more difficult than is necessary to further befuddle us. As John Finn has suggested, in this paradox we could just have well started with a simpler problem. Suppose Ali and Baba know that I am going to give then either an envelope with \$5 or one with \$10 and I am going to toss a coin to decide which to give to Ali, and then give the other to Baba. Then Ali can argue that Baba has $2x$ with probability 1/2 and $x/2$ with probability 1/2. This leads Ali to the same conclusion as before. But now it is clear that this is nonsense, since if Ali has the envelope containing \$5, Baba cannot possibly have half of this, namely \$2.50, since that was not even one of the choices. Similarly, if Ali has \$10, Baba cannot have twice as much, namely \$20. In fact, in this simpler problem the possibly outcomes are given by the tree diagram in the following figure. From the diagram, it is clear that neither is made better off by switching.

In the above example, Ali's reasoning is incorrect because he infers that if the amount in his envelope is x, then the probability that his envelope contains the

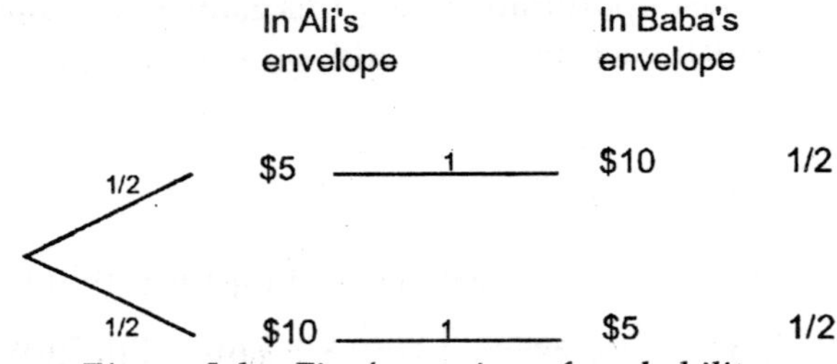

Figure: John Finn's version of probability.

smaller amount is 1/2, and the probability that her envelope contains the larger amount is also 1/2. In fact, these conditional probabilities depend upon the distribution of the amounts that are placed in the envelopes.

For definiteness, let X denote the positive integer-valued random variable which represents the smaller of the two amounts in the envelopes. Suppose, in addition, that we are given the distribution of X, i.e., for each positive integer x, we are given the value of

$$p_x = P(X = x).$$

(In Finn's example, $p_5 = 1$, and $p_n = 0$ for all other values of n). Then it is easy to calculate the conditional probability that an envelope contains the smaller amount, given that it contains x dollars. The two possible sample points are $(x, x/2)$ and $(x, 2x)$. If x is odd, then the first sample point has probability 0, since $x/2$ is not an integer, so the desired conditional probability is 1 that x is the smaller amount. If x is even, then the two sample points have probabilities $p_{x/2}$ and p_x, respectively, so the conditional probability that x is the smaller amount is

$$\frac{p_x}{p_{x/2}+p_x},$$

which is not necessarily equal to 1/2.

Steven Brams and D. Marc Kilgour study the problem, for different distributions, of whether or not one should switch envelopes, if one's objective is to maximise the long-term average winnings. Let x be the amount in your envelope.

They show that for any distribution of X, there is at least one value of x such that you should switch. They give an example of a distribution for which there is exactly one value of x such that you should switch. Perhaps the most interesting case is a distribution in which you should always switch. We now give this example.

Example: Suppose that we have two envelopes in front of us, and that one envelope contains twice the amount of money as the other (both amounts are positive integers). We are given one of the envelopes, and asked if we would like to switch.

As above, we let X denote the smaller of the two amounts in the envelopes, and let

$$p_x = P(X = x).$$

We are now in a position where we can calculate the long-term average winnings, if we switch. (This long-term average is an example of a probabilistic concept known as expectation). Given that one of the two sample points has occurred, the probability that it is the point (x, x/2) is

$$\frac{p_x/2}{p_{x/2}+p_x},$$

and the probability that it is the point $(x, 2x)$ is

$$\frac{p_x}{p_{x/2}+p_x}.$$

Thus, if we switch, our long-term average winnings are

$$\frac{p_{x/2}}{p_{x/2}+p_x}\frac{x}{2}+\frac{p_x}{p_{x/2}+p_x}2x.$$

If this is greater than x, then it pays in the long run for us to switch. Some routine algebra shows that the above expression is greater than x if and only if

$$\frac{p_{x/2}}{p_{x/2}+p_x} < \frac{2}{3}. \tag{4.6}$$

It is interesting to consider whether there is a distribution on the positive integers such that the inequality 4.6 is true for all even values of x. Brams and Kilgour give the following example.

We define p_x as follows:

$$p_x = \begin{cases} \frac{1}{3}\left(\frac{2}{3}\right)^{k-1}, & \text{if } x = 2^k, \\ 0, & \text{otherwise}. \end{cases}$$

It is easy to calculate that for all relevant values of x, we have

$$\frac{p_{x/2}}{p_{x/2}+p_x} = \frac{3}{5},$$

which means that the inequality 4.6 is always true.

So far, we have been able to resolve paradoxes by clearly stating the assumptions being made and by precisely stating the models being used. We end this section by describing a paradox which we cannot resolve.

Example: Suppose that we have two envelopes in front of us, and we are told that the envelopes contain X and Y dollars, respectively, where X and Y are different positive integers. We randomly choose one of the envelopes, and we open it, revealing X, say. Is it possible to determine, with probability greater than 1/2, whether X is the smaller of the two dollar amounts?

Even if we have no knowledge of the joint distribution of X and Y, the surprising answer is yes! Here's how to do it. Toss a fair coin until the first time that heads turns up. Let Z denote the number of tosses required plus 1/2. If Z > X, then we say that X is the smaller of the two amounts, and if Z < X, then we say that X is the larger of the two amounts.

First, if Z lies between X and Y, then we are sure to be correct. Since X and Y are unequal, Z lies between them with positive probability. Second, if Z is not between X and Y, then Z is either greater than both X and Y, or is less than both X and Y. In either case, X is the smaller of the two amounts with probability 1/2, by symmetry considerations (remember, we chose the envelope at random). Thus, the probability that we are correct is greater than 1/2.

Exercises

1. One of the first conditional probability paradoxes was provided by Bertrand. It is called the Box Paradox . A cabinet has three drawers. In the first drawer there are two gold balls, in the second drawer there are two silver balls, and in the third drawer there is one silver and one gold ball. A drawer is picked at random and a ball chosen at random from the two balls in the drawer. Given that a gold ball was drawn, what is the probability that the drawer with the two gold balls was chosen?
2. The following problem is called the two aces problem . This problem, dating back to 1936, has been attributed to the English mathematician J. H. C. Whitehead. This problem was also submitted to Marilyn vos Savant by the master of mathematical puzzles Martin Gardner, who remarks that it is one of his favourites.

   A bridge hand has been dealt, i.e. thirteen cards are dealt to each player. Given that your partner has at least one ace, what is the probability that he has at least two aces? Given that your partner has the ace of hearts, what is the probability that he has at least two aces? Answer these questions for a version of bridge in which there are eight cards, namely four aces and four Kings, and each player is dealt two cards. (The reader may wish to solve the problem with a 52-card deck).

3. In the preceding exercise, it is natural to ask "How do we get the information that the given hand has an ace?" Gridgeman considers two different ways that we might get this information. (Again, assume the deck consists of eight cards).

   (a) Assume that the person holding the hand is asked to "Name an ace in your hand" and answers "The ace of hearts." What is the probability that he has a second ace?

   (b) Suppose the person holding the hand is asked the more direct question "Do you have the ace of hearts?" and the answer is yes. What is the probability that he has a second ace?

4. Using the notation introduced earlier, show that in the example of Brams and Kilgour, if x is a positive power of 2, then

$$\frac{p_{x/2}}{p_{x/2}+p_x} = \frac{3}{5}.$$

5. Using the notation introduced earlier, let

$$p_x = \begin{cases} \frac{2}{3}\left(\frac{1}{3}\right)^k, & \text{if } x = 2^k, \\ 0, & \text{otherwise}. \end{cases}$$

   Show that there is exactly one value of x such that if your envelope contains x, then you should switch.

6. Suppose that we are the declarer in a hand of bridge, and we have the King, 9, 8, 7, and 2 of a certain suit, while the dummy has the ace, 10, 5, and 4 of the same suit. Suppose that we want to play this suit in such a way as to maximise the probabilityof having no losers in the suit. We begin by leading the 2 to the ace, and we note that the Queen drops on our left. We then lead the 10 from the dummy, and our right-hand opponent plays the six (after playing the three on the first round). Should we finesse or play for the drop?

Combinatorics

Permutations

Many problems in probability theory require that we count the number of ways that a particular event can occur. For this, we study the topics of permutations and combinations.

Before discussing permutations, it is useful to introduce a general counting technique that will enable us to solve a variety of counting problems, including the problem of counting the number of possible permutations of n objects.

Counting Problems

Consider an experiment that takes place in several stages and is such that the number of outcomes m at the nth stage is independent of the outcomes of the previous stages. The number m may be different for different stages. We want to count the number of ways that the entire experiment can be carried out.

Example: You are eating at Emile's restaurant and the waiter informs you that you have (a) two choices for appetisers: soup or juice; (b) three for the main course: a meat, fish, or vegetable dish; and (c) two for dessert: ice cream or cake. How many possible choices do you have for your complete meal? We illustrate the possible meals by a tree diagram shown in the following figure.

Your menu is decided in three stages—at each stage the number of possible choices does not depend on what is chosen in the previous stages: two choices at the first stage, three at the second, and two at the third. From the tree diagram we see that the total number of choices is the product of the number of choices at each stage. In this examples, we have $2 \cdot 3 \cdot 2 = 12$ possible menus. Our menu example is an example of the following general counting technique.

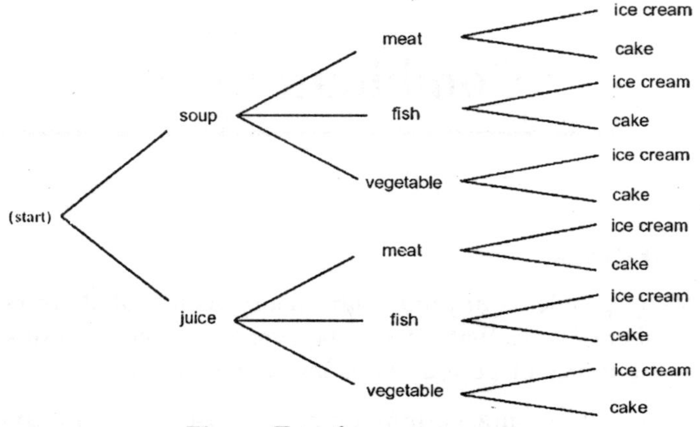

Figure: Tree for your menu.

A Counting Technique

A task is to be carried out in a sequence of r stages. There are n_1 ways to carry out the first stage; for each of these n_1 ways, there are n_2 ways to carry out the second stage; for each of these n_2 ways, there are n_3 ways to carry out the third stage, and so forth. Then the total number of ways in which the entire task can be accomplished is given by the product $N = n_1 \cdot n_2 \cdot \ldots \cdot n_r$.

Tree Diagrams

It will often be useful to use a tree diagram when studying probabilities of events relating to experiments that take place in stages and for which we are given the probabilities for the outcomes at each stage. For example, assume that the owner of Emile's restaurant has observed that 80 per cent of his customers choose the soup for an appetiser and 20 per cent choose juice. Of those who choose soup, 50 per cent choose meat, 30 per cent choose fish, and 20 per cent choose the vegetable dish. Of those who choose juice for an appetiser, 30 per cent choose meat, 40 per cent choose fish, and 30 per cent choose the vegetable dish. We can use this to estimate the probabilities at the first two stages as indicated on the tree diagram of the following figure.

We choose for our sample space the set Ω of all possible paths $w = w_1, w_2, \ldots, w_6$ through the tree. How should we assign our probability distribution? For example, what probability should we assign to the customer choosing soup and then the meat? If 8/10 of the customers choose soup and then 1/2 of these choose meat, a proportion $8/10 \cdot 1/2 = 4/10$ of the customers choose soup and then meat. This suggests choosing our probability distribution for each path through the tree to be the product of the probabilities at each of the stages along the path. This results in the probability measure for the sample points w indicated in the following figure. (Note that $m(w_1) + \cdots + m(w_6) = 1$.) From this, we see, for example, that the probability

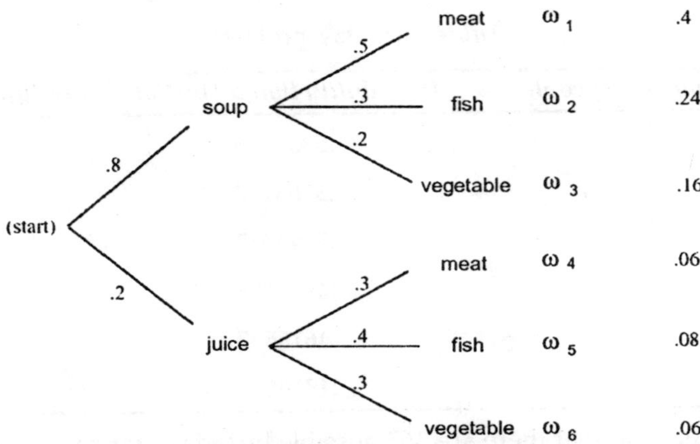

Figure: Two-stage probability assignment.

that a customer chooses meat is $m(w_1) + m(w_4) = .46$.

We return now to more counting problems.

Example: We can show that there are at least two people in Columbus, Ohio, who have the same three initials. Assuming that each person has three initials, there are 26 possibilities for a person's first initial, 26 for the second, and 26 for the third. Therefore, there are 26^3 = 17,576 possible sets of initials. This number is smaller than the number of people living in Columbus, Ohio; hence, there must be at least two people with the same three initials.

We consider next the celebrated birthday problem—often used to show that naive intuition cannot always be trusted in probability.

Birthday Problem

Example: How many people do we need to have in a room to make it a favourable bet (probability of success greater than 1/2) that two people in the room will have the same birthday?

Since there are 365 possible birthdays, it is tempting to guess that we would need about 1/2 this number, or 183. You would surely win this bet. In fact, the number required for a favourable bet is only 23. To show this, we find the probability pr that, in a room with r people, there is no duplication of birthdays; we will have a favourable bet if this probability is less than one half.

Table: Birthday problem.

Number of people	*Probability that all birthdays are different*
20	.5885616
21	.5563117
22	.5243047
23	.4927028
24	.4616557
25	.4313003

Assume that there are 365 possible birthdays for each person (we ignore leap years). Order the people from 1 to r. For a sample point w, we choose a possible sequence of length r of birthdays each chosen as one of the 365 possible dates. There are 365 possibilities for the first element of the sequence, and for each of these choices, there are 365 for the second, and so forth, making

365^r possible sequences of birthdays. We must find the number of these sequences that have no duplication of birthdays.

For such a sequence, we can choose any of the 365 days for the first element, then any of the remaining 364 for the second, 363 for the third, and so forth, until we make r choices. For the rth choice, there will be $365 - r + 1$ possibilities. Hence, the total number of sequences with no duplications is

$$365 \cdot 364 \cdot 363 \cdot \ldots \cdot (365 - r + 1).$$

Thus, assuming that each sequence is equally likely,

$$p_r = \frac{365 \cdot 364 \cdot \ldots \cdot (365 - r + 1)}{365^r}$$

We denote the product

$$(n)(n - 1) \cdots (n - r + 1)$$

by $(n)_r$ (read "*n* down *r*," or "*n* lower *r*"). Thus,

$$p_r = \frac{(365)_r}{(365)^r} \cdot$$

The programme *Birthday* carries out this computation and prints the probabilities for r = 20 to 25. Running this programme, we get the results shown in the previous table. As we asserted above, the probability for no duplication changes from greater than one half to less than one half as we move from 22 to 23 people.

To see how unlikely it is that we would lose our bet for larger numbers of people, we have run the programme again, printing out values from $r = 10$ to $r = 100$ in steps of 10. We see that in a room of 40 people the odds already heavily favour a duplication, and in a room of 100 the odds are overwhelmingly in favour of a duplication.

We have assumed that birthdays are equally likely to fall on any particular day. Statistical evidence suggests that this is not true. However, it is intuitively clear (but not easy to prove) that this makes it even more likely to have a duplication with a group of 23 people.

Permutations

Table: Birthday problem.

Number of people	*Probability that all birthdays are different*
10	.8830518
20	.5885616
30	.2936838
40	.1087682
50	.0296264
60	.0058773
70	.0008404
80	.0000857
90	.0000062
100	.0000003

We now turn to the topic of permutations.

Definition: Let A be any finite set. A permutation of A is a one-to-one mapping of A onto itself.

To specify a particular permutation, we list the elements of A and, under them, show where each element is sent by the one-to-one mapping. For example, if $A = \{a, b, c\}$, a possible permutation σ would be

$$\sigma = \begin{pmatrix} a & b & c \\ b & c & a \end{pmatrix}.$$

By the permutation σ, a is sent to b, b is sent to c, and c is sent to a. The condition that the mapping be one-to-one means that no two elements of A are sent, by the mapping, into the same element of A.

We can put the elements of our set in some order and rename them 1, 2,..., n. Then, a typical permutation of the set $A = \{a_1, a_2, a_3, a_4\}$ can be written in the form

$$\sigma = \begin{pmatrix} 1 & 2 & 3 & 4 \\ 2 & 1 & 4 & 3 \end{pmatrix},$$

indicating that a_1 went to a_2, a_2 to a_1, a_3 to a_4, and a_4 to a_3.

If we always choose the top row to be 1 2 3 4 then, to prescribe the permutation, we need only give the bottom row, with the understanding that this tells us where 1 goes, 2 goes, and so forth, under the mapping. When this is done, the permutation is often called a rearrangement of the n objects 1, 2, 3,..., n. For example, all possible permutations, or rearrangements, of the numbers $A = \{1, 2, 3\}$ are:

123, 132, 213, 231, 312, 321.

It is an easy matter to count the number of possible permutations of n objects. By our general counting principle, there are n ways to assign the first element, for

Table: Values of the factorial function.

n	n!
0	1
1	1
2	2
3	6
4	24
5	120
6	720
7	5040
8	40320
9	362880
10	3628800

each of these, we have $n - 1$ ways to assign the second object, $n - 2$ for the third, and so forth. This proves the following theorem.

Theorem (a): The total number of permutations of a set A of n elements is given by n. (n – 1). (n – 2)·....·1.

It is sometimes helpful to consider orderings of subsets of a given set. This prompts the following definition.

Definition: Let A be an n-element set, and let k be an integer between 0 and n. Then a k-permutation of A is an ordered listing of a subset of A of size k.

Using the same techniques as in the last theorem, the following result is easily proved.

Theorem (b): The total number of k-permutations of a set A of n elements is given by $n\cdot(n-1)\cdot(n-2)\cdot\ldots\cdot(n-k+1)$.

Factorials

The number given in Theorem (a) is called n factorial, and is denoted by $n!$. The expression 0! is defined to be 1 to make certain formulas come out simpler. The first few values of this function are shown in table above. The reader will note that this function grows very rapidly.

The expression $n!$ will enter into many of our calculations, and we shall need to have some estimate of its magnitude when n is large. It is clearly not practical to make exact calculations in this case. We shall instead use a result called Stirling's formula. Before stating this formula, we need a definition.

Table: Stirling approximations to the factorial function.

n	*n!*	*Approximation*	*Ratio*
1	1	.922	1.084
2	2	1.919	1.042
3	6	5.836	1.028
4	24	23.506	1.021
5	120	118.019	1.016
6	720	710.078	1.013

7	5040	4980.396	1.011
8	40320	39902.395	1.010
9	362880	359536.873	1.009
10	3628800	3598696.619	1.008

Definition: Let a_n and b_n be two sequences of numbers. We say that a_n is asymptotically equal to b_n, and write $a_n \sim b_n$, if

$$\lim_{x \to \infty} \frac{a_n}{b_n} = 1$$

Example: If $a_n = n + \sqrt{n}$ and $b_n = n$ then, since $a_n/b_n = 1 + 1/\sqrt{n}$ and this ratio tends to 1 as n tends to infinity, we have $a_n \sim b_n$.

Theorem (c): *(Stirling's Formula)* The sequence $n!$ is asymptotically equal to

$$n^n e^{-n} \sqrt{2\pi n}\,.$$

The proof of Stirling's formula may be found in most analysis texts. Let us verify this approximation by using the computer. The programme Stirling Approximations prints $n!$, the Stirling approximation, and, finally, the ratio of these two numbers. Sample output of this programme is shown in the previous table. Note that, while the ratio of the numbers is getting closer to 1, the difference between the exact value and the approximation is increasing, and indeed, this difference will tend to infinity as n tends to infinity, even though the ratio tends to 1. (This was also true in our Example where $n + \sqrt{n} \sim n$, but the difference is $\sqrt{n}$.)

Generating Random Permutations

We now consider the question of generating a random permutation of the integers between 1 and n. Consider the following experiment. We start with a deck of n cards, labelled 1 through n. We choose a random card out of the deck, note its label, and put the card aside. We repeat this process until all n cards have been chosen. It is clear that each permutation of the integers from 1 to n can occur as a sequence

Table: Fixed Point Distributions.

Number of fixed points	*Fraction of permutations*		
	$n = 10$	$n = 20$	$n = 30$
0	.362	.370	.358
1	.368	.396	.358
2	.202	.164	.192
3	.052	.060	.070
4	.012	.008	.020
5	.004	.002	.002
Average number of fixed points	*1.000*	*1.000*	*1.000*

of labels in this experiment, and that each sequence of labels is equally likely to occur. In our implementations of the computer algorithms, the above procedure is called *RandomPermutation.*

Fixed Points

There are many interesting problems that relate to properties of a permutation chosen at random from the set of all permutations of a given finite set. For example, since a permutation is a one-to-one mapping of the set onto itself, it is interesting to ask how many points are mapped onto themselves. We call such points fixed points of the mapping.

Let $p_k(n)$ be the probability that a random permutation of the set $\{1, 2,..., n\}$ has exactly k fixed points. We will attempt to learn something about these probabilities using simulation. The programme FixedPoints uses the procedure RandomPermutation to generate random permutations and count fixed points.

The programme prints the proportion of times that there are k fixed points as well as the average number of fixed points. The results of this programme for 500 simulations for the cases $n = 10$, 20, and 30 are shown in the previous table. Notice the rather surprising fact that our estimates for the probabilities do not seem to depend very heavily on the number of elements in the permutation. For example, the probability that there are no fixed points, when $n = 10, 20$, or 30 is estimated to be between .35 and .37. We shall see later that for $n \geq 10$ the exact probabilities $p_n(0)$ are, to six decimal place accuracy, equal to $1/e \approx .367879$.

Thus, for all practical purposes, after $n = 10$ the probability that a random permutation of the set $\{1, 2, \ldots, n\}$ has no fixed points does not depend upon n. These simulations also suggest that the average number of fixed points is close to 1. It can be shown that the average is exactly equal to 1 for all n.

More picturesque versions of the fixed-point problem are: You have arranged the books on your book shelf in alphabetical order by author and they get returned to your shelf at random; what is the probability that exactly k of the books end up in their correct position? (The library problem.) In a restaurant, n hats are checked and they are hopelessly scrambled; what is the probability that no one gets his own hat back? (The hat check problem).

Table: Snowfall in Hanover.

Date	***Snowfall in inches***
1974	75
1975	88
1976	72
1977	110
1978	85
1979	30
1980	55
1981	86
1982	51
1983	64

Table: Ranking of total snowfall.

Year	1	2	3	4	5	6	7	8	9	10
Ranking	6	9	5	10	7	1	3	8	2	4

Records

Here is another interesting probability problem that involves permutations. Estimates for the amount of measured snow in inches in Hanover, New Hampshire, in the ten years from 1974 to 1983 are shown in the previous table. Suppose we have started keeping records in 1974. Then our first year's snowfall could be considered a record snowfall starting from this year. A new record was established in 1975; the next record was established in 1977, and there were no new records established after this year.

Thus, in this ten-year period, there were three records established: 1974, 1975, and 1977. The question that we ask is: How many records should we expect to be established in such a ten-year period? We can count the number of records in terms of a permutation as follows: We number the years from 1 to 10. The actual amounts of snowfall are not important but their relative sizes are. We can, therefore, change the numbers measuring snowfalls to numbers 1 to 10 by replacing the smallest number by 1, the next smallest by 2, and so forth. (We assume that there are no ties.) For our example, we obtain the data shown in the previous table.

This gives us a permutation of the numbers from 1 to 10 and, from this permutation, we can read off the records; they are in years 1, 2, and 4. Thus, we can define records for a permutation as follows:

Definition: Let σ be a permutation of the set {1, 2,..., n}. Then i is a record of σ if either $i = 1$ or $\sigma(j) < \sigma(i)$ for every $j = 1;..., i - 1$.

Now if we regard all rankings of snowfalls over an n-year period to be equally likely (and allow no ties), we can estimate the probability that there will be k records in n years as well as the average number of records by simulation.

We have written a programme *Records* that counts the number of records in randomly chosen permutations. We have run this programme for the cases $n = 10, 20, 30$. For $n = 10$, the average number of records is 2.968, for 20 it is 3.656, and for 30 it is 3.960. We see now that the averages increase, but very slowly. We shall see later that the average number is approximately $\log n$. Since $\log 10 = 2.3$, $\log 20 = 3$, and $\log 30 = 3.4$, this is consistent with the results of our simulations.

As remarked earlier, we shall be able to obtain formulas for exact results of certain problems of the above type. However, only minor changes in the problem make this impossible. The power of simulation is that minor changes in a problem do not make the simulation much more difficult.

List of Permutations

Another method to solve problems that is not sensitive to small changes in the problem is to have the computer simply list all possible permutations and count the fraction that have the desired property. The programme *AllPermutations* produces a list of all of the permutations of n. When we try running this programme, we run into a limitation on the use of the computer. The number of permutations of n increases so rapidly that even to list all permutations of 20 objects is impractical.

Historical Remarks

Our basic counting principle stated that if you can do one thing in r ways and for each of these another thing in s ways, then you can do the pair in rs ways. This is such a self-evident result that you might expect that it occurred very early in mathematics. N. L. Biggs suggests that we might trace an example of this principle as follows: First, he relates a popular nursery rhyme dating back to at least 1730:

As I was going to St. Ives,

I met a man with seven wives,

Each wife had seven sacks,

Each sack had seven cats,

Each cat had seven kits.

Kits, cats, sacks and wives,

How many were going to St. Ives?

(You need our principle only if you are not clever enough to realise that you are supposed to answer one, since only the narrator is going to St. Ives; the others are going in the other direction!)

He also gives a problem appearing on one of the oldest surviving mathematical manuscripts of about 1650 BC, roughly translated as:

Houses	7
Cats	49
Mice	343
Wheat	2401
Hekat	16807
	19607

The following interpretation has been suggested: there are seven houses, each with seven cats; each cat kills seven mice; each mouse would have eaten seven heads of wheat, each of which would have produced seven hekat measures of grain. With this interpretation, the table answers the question of how many hekat measures were saved by the cats' actions. It is not clear why the writer of the table wanted to add the numbers together.

One of the earliest uses of factorials occurred in Euclid's proof that there are infinitely many prime numbers. Euclid argued that there must be a prime number between n and $n! + 1$ as follows: $n!$ and $n! + 1$ cannot have common factors. Either $n! + 1$ is prime or it has a proper factor. In the latter case, this factor cannot divide n! and hence must be between n and $n! + 1$. If this factor is not prime, then it has a factor that, by the same argument, must be bigger than n. In this way, we eventually reach a prime bigger than n, and this holds for all n.

The "*n*!" rule for the number of permutations seems to have occurred first in India. Examples have been found as early as 300 BC, and by the eleventh century the general formula seems to have been well known in India and then in the Arab countries.

The *hat check* problem is found in an early probability book written by de Montmort and first printed in 1708. It appears in the form of a game called Treize. In a simplified version of this game considered by de Montmort, one turns over cards numbered 1 to 13, calling out 1, 2,..., 13 as the cards are examined. De Montmort asked for the probability that no card that is turned up agrees with the number called out.

This probability is the same as the probability that a random permutation of 13 elements has no fixed point. De Montmort solved this problem by the use of a recursion relation as follows: let w_n be the number of permutations of *n* elements with no fixed point (such permutations are called derangements). Then $w_1 = 0$ and $w_2 = 1$.

Now assume that $n > 3$ and choose a derangement of the integers between 1 and *n*. Let *k* be the integer in the first position in this derangement. By the definition of derangement, we have $k \neq 1$. There are two possibilities of interest concerning the position of 1 in the derangement: either 1 is in the *k*th position or it is elsewhere. In the first case, the $n - 2$ remaining integers can be positioned in w_{n-2} ways, without resulting in any fixed points. In the second case, we consider the set of integers {1, 2,..., k – 1, *k* + 1,..., n}. The numbers in this set must occupy the positions {2, 3,..., n} so that none of the numbers other than 1 in this set are fixed, and also so that 1 is not in position *k*. The number of ways of achieving this kind of arrangement is just w_{n-1}. Since there are $n - 1$ possible values of *k*, we see that

$$w_n = (n-1)w_{n-1} + (n-1)w_{n-2}$$

for $n \geq 3$. One might conjecture from this last equation that the sequence fw$_n$g grows like the sequence {n!}.

In fact, it is easy to prove by induction that

$$w_n = nw_{n-1} + (-1)^n.$$

Then $p_i = w_i/i!$ satisfies

$$p_i - p_{i-1} = \frac{(-1)^i}{i!}.$$

If we sum from $i = 2$ to n, and use the fact that $p_1 = 0$, we obtain

$$p_n = \frac{1}{2!} - \frac{1}{3!} + ... + \frac{(-1)^n}{n!}$$

This agrees with the first $n + 1$ terms of the expansion for e^x for $x = -1$ and hence for large n is approximately $e^{-1} \approx .368$. David remarks that this was possibly the first use of the exponential function in probability. We shall see another way to derive de Montmort's result later, using a method known as the Inclusion-Exclusion method.

Recently, a related problem appeared in a column of Marilyn vos Savant. Charles Price wrote to ask about his experience playing a certain form of solitaire, sometimes called "frustration solitaire." In this particular game, a deck of cards is shuffled, and then dealt out, one card at a time. As the cards are being dealt, the player counts from 1 to 13, and then starts again at 1. (Thus, each number is counted four times.) If a number that is being counted coincides with the rank of the card that is being turned up, then the player loses the game. Price found that he rarely won and wondered how often he should win. *Vos Savant* remarked that the expected number of matches is 4 so it should be difficult to win the game.

Finding the chance of winning is a harder problem than the one that de Montmort solved because, when one goes through the entire deck, there are different patterns for the matches that might occur. For example, matches may occur for two cards of the same rank, say two aces, or for two different ranks, say a two and a three.

A discussion of this problem can be found in Riordan. In this book, it is shown that as $n \to \infty$, the probability of no matches tends to $1/e^4$.

The original game of Treize is more difficult to analyse than frustration solitaire. The game of Treize is played as follows. One person is chosen as dealer and the others are players. Each player, other than the dealer, puts up a stake. The dealer shuffles the cards and turns them up one at a time calling out, "Ace, two, three,..., King," just as in frustration solitaire. If the dealer goes through the 13 cards without a match, he pays the players an amount equal to their stake, and the deal passes to someone else. If there is a match the dealer collects the players' stakes; the players put up new stakes, and the dealer continues through the deck, calling out, "Ace, two, three,...." If the dealer runs out of cards he reshuffles and continues the count where he left off. He continues until there is a run of 13 without a match and then a new dealer is chosen.

The question at this point is how much money can the dealer expect to win from each player. *De Montmort* found that if each player puts up a stake of 1, say, then the dealer will win approximately .801 from each player.

Peter Doyle calculated the exact amount that the dealer can expect to win. The answer is:

26516072156010218582227607912734182784642120
48213609144671537196208993152311343541724554334912870
5414402992392516076941135000807759178185120
13821768766535631738528745558593672546
32009477403727395572807459384342747
87664965076063990538261189388143513
5473663160170049455072017642788283066
01171079536331427343824779227098352817
53299035988581413688367655833113
24476153310720627474169719301806649
15269870408438391421790790695497603628
52821159014031620212060154912692088082
49133255538826920554278308103685
78188612087582488006809786404381185
828348775425609555506628789271230482
69976017001162335927933082975336421935050
745402689256831938878213014427051979188 2/

3303692913358259222011722071315607111497510114
9831063364072138969878007996

4720470882530338752589223658132301562800562114342729062565897443397165719454122908007086289841306087561302818991167357863623756067184986491353535536221974488902232671011588010162859313519792943872232770333969677979706993347580242367694987366160518403147756156039338025707097071195969641268242455013319879747054693517809383750593488858698672364846950539888686285826099055862710013181506211344070569832147402218515677066720809458658937845943279986870633416181298863049632728725481845887935302449800322425586446741048147720934108061350613503856973048971213063937040515595337315 91.

This is .803 to 3 decimal places. A discussion of this problem and other problems can be found in Doyle *et al.*

The birthday problem does not seem to have a very old history. Problems of this type were first discussed by von Mises. It was made popular in the 1950s by Feller's book.

Stirling presented his formula in his work Methodus Differentialis published in 1730.

$$n! \sim \sqrt{2\pi n}\left(\frac{n}{e}\right)^n$$

This approximation was used by de Moivre in establishing his celebrated central limit theorem. *De Moivre* himself had independently established this approximation, but without identifying the constant π. Having established the approximation

$$\frac{2B}{\sqrt{n}}$$

for the central term of the binomial distribution, where the constant B was determined by an infinite series, *de Moivre* writes:

... my worthy and learned Friend, Mr. James Stirling, who had applied himself after me to that enquiry, found that the Quantity B did denote the Square-root of the Circumference of a Circle whose Radius is Unity, so that if that Circumference be called c the Ratio of the middle Term to the Sum of all Terms will be expressed by $2/\sqrt{nc}$ ....

Exercises

1. Four people are to be arranged in a row to have their picture taken. In how many ways can this be done?
2. An automobile manufacturer has four colours available for automobile exteriors and three for interiors. How many different colour combinations can he produce?
3. In a digital computer, a *bit* is one of the integers {0,1}, and a word is any string of 32 bits. How many different words are possible?
4. What is the probability that at least 2 of the Presidents of the United States have died on the same day of the year? If you bet this has happened, would you win your bet?
5. There are three different routes connecting city A to city B. How many ways can a round trip be made from A to B and back? How many ways if it is desired to take a different route on the way back?
6. In arranging people around a circular table, we take into account their seats relative to each other, not the actual position of any one person. Show that n people can be arranged around a circular table in $(n - 1)!$ ways.
7. Five people get on an elevator that stops at five floors. Assuming that each has an equal probability of going to any one floor, find the probability that they all get off at different floors.
8. A finite set Ω has n elements. Show that if we count the empty set and Ω as subsets, there are 2^n subsets of Ω.

9. A more refined inequality for approximating n! is given by

$$\sqrt{2\pi n}\left(\frac{n}{e}\right)^n e^{1/(12n+1)} < n! < \sqrt{2\pi n}\left(\frac{n}{e}\right)^n e^{1/(12n)}$$

Write a computer programme to illustrate this inequality for $n = 1$ to 9.

10. A deck of ordinary cards is shuffled and 13 cards are dealt. What is the probability that the last card dealt is an ace?

11. There are n applicants for the director of computing. The applicants are interviewed independently by each member of the three-person search committee and ranked from 1 to n. A candidate will be hired if he or she is ranked first by at least two of the three interviewers. Find the probability that a candidate will be accepted if the members of the committee really have no ability at all to judge the candidates and just rank the candidates randomly. In particular, compare this probability for the case of three candidates and the case of ten candidates.

12. A symphony orchestra has in its repertoire 30 Haydn symphonies, 15 modern works, and 9 Beethoven symphonies. Its programme always consists of a Haydn symphony followed by a modern work, and then a Beethoven symphony.
    (a) How many different programmes can it play?
    (b) How many different programmes are there if the three pieces can be played in any order?
    (c) How many different three-piece programmes are there if more than one piece from the same category can be played and they can be played in any order?

13. A certain state has licence plates showing three numbers and three letters.

    How many different licence plates are possible:
    (a) if the numbers must come before the letters?
    (b) if there is no restriction on where the letters and numbers appear?

14. The door on the computer centre has a lock which has five buttons numbered from 1 to 5. The combination of numbers that opens the lock is a sequence of five numbers and is reset every week.

    (a) How many combinations are possible if every button must be used once?

    (b) Assume that the lock can also have combinations that require you to push two buttons simultaneously and then the other three one at a time.

    How many more combinations does this permit?

15. A computing centre has 3 processors that receive n jobs, with the jobs assigned to the processors purely at random so that all of the 3^n possible assignments are equally likely. Find the probability that exactly one processor has no jobs.

16. Prove that at least two people in Atlanta, Georgia, have the same initials, assuming no one has more than four initials.

17. Find a formula for the probability that among a set of n people, at least two have their birthdays in the same month of the year (assuming the months are equally likely for birthdays).

18. Consider the problem of finding the probability of more than one coincidence of birthdays in a group of n people. These include, for example, three people with the same birthday, or two pairs of people with the same birthday, or larger coincidences. Show how you could compute this probability, and write a computer programme to carry out this computation. Use your programme to find the smallest number of people for which it would be a favourable bet that there would be more than one coincidence of birthdays.

19. Suppose that on planet Zorg, a year has n days, and that the lifeforms there are equally likely to have hatched on any day of the year. We would like to estimate d, which is the minimum number of lifeforms needed so that the probability of at least two sharing a birthday exceeds 1/2.

(a) In Example, it was shown that in a set of d lifeforms, the probability that no two life forms share a birthday is

$$\frac{(n)_d}{n^d},$$

where $(n)_d = (n)(n-1)\ldots(n-d+1)$. Thus, we would like to set this equal to 1/2 and solve for d.

(b) Using Stirling's Formula, show that

$$\frac{(n)_d}{n^d} \sim \left(1+\frac{d}{n-d}\right)^{n-d+1/2} e^{-d}.$$

(c) Now take the logarithm of the right-hand expression, and use the fact that for small values of x, we have

$$\log(1+x) \sim x - \frac{x^2}{2}$$

(We are implicitly using the fact that d is of smaller order of magnitude than n. We will also use this fact in part (d)).

(d) Set the expression found in part (c) equal to ¡ log(2), and solve for d as a function of n, thereby showing that

$$d \sim \sqrt{2(\log 2)n}$$

Hint: If all three summands in the expression found in part (b) are used, one obtains a cubic equation in d. If the smallest of the three terms is thrown away, one obtains a quadratic equation in d.

(e) Use a computer to calculate the exact values of d for various values of n. Compare these values with the approximate values obtained by using the answer to part (d).

20. At a mathematical conference, ten participants are randomly seated around a circular table for meals. Using simulation, estimate the probability that no two people sit next to each other at both lunch and dinner. Can you make an intelligent conjecture for the case of n participants when n is large?

21. Modify the programme *AllPermutations* to count the number of permutations of n objects that have exactly j fixed points

for $j = 0, 1, 2,..., n$. Run your programme for $n = 2$ to 6. Make a conjecture for the relation between the number that have 0 fixed points and the number that have exactly 1 fixed point. A proof of the correct conjecture can be found in Wilf.

22. Mr. Wimply Dimple, one of London's most prestigious watch makers, has come to Sherlock Holmes in a panic, having discovered that someone has been producing and selling crude counterfeits of his best selling watch. The 16 counterfeits so far discovered bear stamped numbers, all of which fall between 1 and 56, and Dimple is anxious to know the extent of the forger's work. All present agree that it seems reasonable to assume that the counterfeits, thus far, produced bear consecutive numbers from 1 to whatever the total number is.

    "Chin up, Dimple," opines Dr. Watson."I shouldn't worry overly much if I were you; the Maximum Likelihood Principle, which estimates the total number as precisely that which gives the highest probability for the series of numbers found, suggests that we guess 56 itself as the total. Thus, your forgers are not a big operation, and we shall have them safely behind bars before your business suffers significantly." "Stuff, nonsense, and bother your fancy principles, Watson," counters Holmes. "Anyone can see that, of course, there must be quite a few more than 56 watches—why the odds of our having discovered precisely the highest numbered watch made are laughably negligible. A much better guess would be twice 56."

    (a) Show that Watson is correct that the Maximum Likelihood Principle gives 56.

    (b) Write a computer programme to compare Holmes's and Watson's guessing strategies as follows: fix a total N and choose 16 integers randomly between 1 and N. Let m denote the largest of these. Then Watson's guess for N is m, while Holmes's is 2 m. See which of these is closer to N. Repeat this experiment (with N still fixed) a hundred or more times, and determine the proportion of times that each comes closer. Whose seems to be the better strategy?

23. Barbara Smith is interviewing candidates to be her secretary. As she interviews the candidates, she can determine the relative rank of the candidates but not the true rank. Thus, if there are six candidates and their true rank is 6, 1, 4, 2, 3, 5, (where 1 is best) then after she had interviewed the first three candidates she would rank them 3, 1, 2. As she interviews each candidate, she must either accept or reject the candidate. If she does not accept the candidate after the interview, the candidate is lost to her. She wants to decide on a strategy for deciding when to stop and accept a candidate that will maximise the probability of getting the best candidate. Assume that there are n candidates and they arrive in a random rank order.

    (a) What is the probability that Barbara gets the best candidate if she inter views all of the candidates? What is it if she chooses the first candidate?

    (b) Assume that Barbara decides to interview the first half of the candidates and then continue interviewing until getting a candidate better than any candidate seen so far. Show that she has a better than 25 per cent chance of ending up with the best candidate.

24. For the task described in Exercise 23, it can be shown that the best strategy is to pass over the first k – 1 candidates where k is the smallest integer for which

$$\frac{1}{k}+\frac{1}{k+1}+\ldots+\frac{1}{n-1}\leq 1.$$

Using this strategy, the probability of getting the best candidate is approximately 1/e = .368. Write a programme to simulate Barbara Smith's interviewing if she uses this optimal strategy, using $n = 10$, and see if you can verify that the probability of success is approximately 1/e.

Combinations

Having mastered permutations, we now consider combinations. Let U be a set with n elements; we want to count the number of distinct subsets of the set U that have exactly j elements. The empty set and the set U are considered to be subsets of U. The empty set is usually denoted by ϕ.

Example Let $U = \{a, b, c\}$. The subsets of U are:

ϕ, $\{a\}$, $\{b\}$, $\{c\}$, $\{a, b\}$, $\{a, c\}$, $\{b, c\}$, $\{a, b, c\}$.

Binomial Coefficients

The number of distinct subsets with j elements that can be chosen from a set with n elements is denoted by $\binom{n}{j}$, and is pronounced "n choose j." The number $\binom{n}{j}$ is called a binomial coefficient. This terminology comes from an application to algebra.

In the above example, there is one subset with no elements, three subsets with exactly 1 element, three subsets with exactly 2 elements, and one subset with exactly 3 elements. Thus, $\binom{3}{0} = 1$, $\binom{3}{1} = 3$, $\binom{3}{2} = 3$, and $\binom{3}{3} = 1$. Note that there are $2^3 = 8$ subsets in all. (We have already seen that a set with n elements has 2^n subsets). It follows that

$$\binom{3}{0}+\binom{3}{1}+\binom{3}{2}+\binom{3}{3} = 2^3 = 8,$$

$$\binom{n}{0}=\binom{n}{n}=1.$$

Assume that $n > 0$. Then, since there is only one way to choose a set with no elements and only way to choose a set with n elements, the remaining values of $\binom{n}{j}$ are determined by the following recurrence relation:

Theorem (d): For integers n and j, with $0 < j < n$, the binomial coefficients satisfy:

$$\binom{n}{j} = \binom{n-1}{j} + \binom{n-1}{j-1} \qquad 5.1$$

Proof: We wish to choose a subset of j elements. Choose an element u of U. Assume first that we do not want u in the subset. Then, we must choose the j elements from a set of $n - 1$ elements U; this can be done in $\binom{n-1}{j}$ ways. On the other hand, assume that we do want u in the subset. Then we must choose the other $j - 1$ elements from the remaining $n - 1$ elements of U; this can be done in $\binom{n-1}{j-1}$ ways. Since u is either in our subset or not, the number of ways that we can choose a subset of j elements is the sum of the number of subsets of j elements which have u as a member and the number which do not—this is what the above equation states.

The binomial coefficient $\binom{n}{j}$ is defined to be 0, if $j < 0$ or if $j > n$. With this definition, the restrictions on j in Theorem given above are unnecessary.

j = 0	1	2	3	4	5	6	7	8	9	10
1										
1	1									
1	2	1								
1	3	3	1							
1	4	6	4	1						
1	5	10	10	5	1					
1	6	15	20	15	6	1				
1	7	21	35	35	21	7	1			
1	8	28	56	70	56	28	8	1		
1	9	36	84	126	126	84	36	9	1	
1	10	45	120	210	252	210	120	45	10	1

Figure: Pascal's triangle.

Pascal's Triangle

The relation 5.1, together with the knowledge that

$$\binom{n}{0} = \binom{n}{n} = 1,$$

determines completely the numbers $\binom{n}{j}$. We can use these relations to determine the famous triangle of Pascal, which exhibits all these numbers in matrix form.

The nth row of this triangle has the entries $\binom{n}{0}, \binom{n}{1}, \ldots, \binom{n}{n}$. We know that the first and last of these numbers are 1. The remaining numbers are determined by the recurrence relation equation (5.1); that is, the entry $\binom{n}{j}$ for $0 < j < n$ in the nth row of Pascal's triangle is the sum of the entry immediately above and the one immediately to its left in the $(n - 1)$st row. For example, $\binom{5}{2} = 6 + 4 = 10$.

This algorithm for constructing Pascal's triangle can be used to write a computer programme to compute the binomial coefficients. You are asked to do this in Exercise 4.

While Pascal's triangle provides a way to construct recursively the binomial coefficients, it is also possible to give a formula for $\binom{n}{j}$.

Theorem (e): The binomial coefficients are given by the formula

$$\binom{n}{j} = \frac{(n)_j}{j!}. \tag{5.2}$$

Proof: Each subset of size j of a set of size n can be ordered in $j!$ ways. Each of these orderings is a j -permutation of the set of size n. The number of j -permutations is $(n)_j$, so the number of subsets of size j is

$$\frac{(n)_j}{j!}.$$

This completes the proof.

The above formula can be rewritten in the form

$$\binom{n}{j} = \frac{n!}{j!(n-j)!}.$$

This immediately shows that

$$\binom{n}{j} = \binom{n}{n-j}.$$

When using equation (5.2) in the calculation of $\binom{n}{j}$, if one alternates the multiplications and divisions, then all of the intermediate values in the calculation are integers. Furthermore, none of these intermediate values exceed the final value.

Another point that should be made concerning equation (5.2) is that if it is used to define the binomial coefficients, then it is no longer necessary to require n to be a positive integer. The variable j must still be a non-negative integer under this definition. This idea is useful when extending the Binomial Theorem to general exponents. [The Binomial Theorem for non-negative integer exponents is given below as Theorem (g)].

Poker Hands

Example: Poker players sometimes wonder why a four of a kind beats a full house. A poker hand is a random subset of 5 elements from a deck of 52 cards. A hand has four of a kind if it has four cards with the same value—for example, four sixes or four Kings. It is a full house if it has three of one value and two of a second—for example, three twos and two Queens. Let us see which hand is more likely. How many hands have four of a kind? There are 13 ways that we can specify the value for the four cards. For each of these, there are 48 possibilities for the fifth card. Thus, the number of four-of-a-kind hands is 13.48=624. Since the total number of possible hands is $\binom{52}{5}$=2598960, the probability of a hand with four of a kind is 624/2598960 = .00024.

Now consider the case of a full house; how many such hands are there? There are 13 choices for the value which occurs three times; for each of these there are $\binom{4}{3} = 4$ choices for the particular three cards of this value that are in the hand. Having picked these three cards, there are 12 possibilities for the value which occurs twice; for each of these there are $\binom{4}{2} = 6$ possibilities for the particular pair of this value. Thus, the number of full houses is $13 \cdot 4 \cdot 12 \cdot 6 = 3744$, and the probability of obtaining a hand with a full house is 3744/2598960 = .0014. Thus, while both types of hands are unlikely, you are six times more likely to obtain a full house than four of a kind.

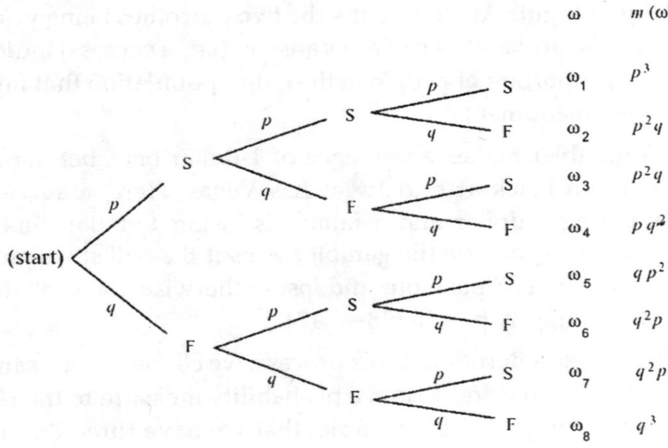

Figure: Tree diagram of three Bernoulli trials.

Bernoulli Trials

Our principal use of the binomial coefficients will occur in the study of one of the important chance processes called *Bernoulli trials.*

Definition: A Bernoulli trials process is a sequence of n chance experiments such that

1. Each experiment has two possible outcomes, which we may call success and failure.
2. The probability p of success on each experiment is the same for each experiment, and this probability is not affected by any knowledge of the previous outcomes. The probability q of failure is given by $q = 1 - p$.

Example: The following are Bernoulli trials processes:

1. A coin is tossed ten times. The two possible outcomes are heads and tails. The probability of heads on any one toss is 1/2.

2. An opinion poll is carried out by asking 1000 people, randomly chosen from the population, if they favour the Equal Rights Amendment—the two outcomes being yes and no. The probability p of a yes answer (i.e., a success) indicates the proportion of people in the entire population that favour this amendment.

3. A gambler makes a sequence of 1-dollar bets, betting each time on black at roulette at Las Vegas. Here, a success is winning 1 dollar and a failure is losing 1 dollar. Since in American roulette the gambler wins if the ball stops on one of 18 out of 38 positions and loses otherwise, the probability of winning is p = 18/38 = .474.

To analyse a Bernoulli trials process, we choose as our sample space a binary tree and assign a probability measure to the paths in this tree. Suppose, for example, that we have three Bernoulli trials. The possible outcomes are indicated in the tree diagram shown in the previous figure. We define X to be the random variable which represents the outcome of the process, i.e., an ordered triple of S's and F's. The probabilities assigned to the branches of the tree represent the probability for each individual trial. Let the outcome of the ith trial be denoted by the random variable X_i, with distribution function m_i. Since we have assumed that outcomes on any one trial do not affect those on another, we assign the same probabilities at each level of the tree. An outcome w for the entire experiment will be a path through the tree. For example, w_3 represents the outcomes SFS. Our frequency interpretation of probability would lead us to expect a fraction p of successes on the first experiment; of these, a fraction q of failures on the second; and, of these, a fraction p of successes on the third experiment. This suggests assigning probability pqp to the outcome w_3.

More generally, we assign a distribution function m(w) for paths w by defining m(w) to be the product of the branch probabilities along the path w. Thus, the probability that the three events S on the first trial, F on the second trial, and S on the third trial occur is the product of the probabilities for the individual events. This means that the events involved are independent in the sense that the knowledge of one event does not affect our prediction for the occurrences of the other events.

Binomial Probabilities

We shall be particularly interested in the probability that in n Bernoulli trials there are exactly j successes. We denote this probability by $b(n, p, j)$. Let us calculate the particular value $b(3, p, 2)$ from our tree measure. We see that there are three paths which have exactly two successes and one failure, namely w_2, w_3, and w_5. Each of these paths has the same probability p^2q. Thus, $b(3, p, 2) = 3p^2q$. Considering all possible numbers of successes we have

$$1b(3, p, 0) = q^3,$$
$$b(3, p, 1) = 3pq^2,$$
$$b(3, p, 2) = 3p^2q,$$
$$b(3, p, 3) = p^3.$$

We can, in the same manner, carry out a tree measure for n experiments and determine $b(n, p, j)$ for the general case of n Bernoulli trials.

Theorem (f): Given n Bernoulli trials with probability p of success on each experiment, the probability of exactly j successes is

$$b(n,p,j) = \binom{n}{j} p^j q^{n-j}$$

where $q = 1 - p$.

Proof: We construct a tree measure as described above. We want to find the sum of the probabilities for all paths which have exactly j successes and $n - j$ failures. Each such path is assigned a probability $p^j q^{n-j}$. How many such paths are there?

To specify a path, we have to pick, from the n possible trials, a subset of j to be successes, with the remaining $n - j$ outcomes being failures. We can do this in $\binom{n}{j}$ ways. Thus, the sum of the probabilities is

$$b(n,p,j) = \binom{n}{j} p^j q^{n-j}.$$

Example: A fair coin is tossed six times. What is the probability that exactly three heads turn up? The answer is:

$$b(6, .5, 3) = \binom{6}{3}\left(\frac{1}{2}\right)^3\left(\frac{1}{2}\right)^3 = 20 \cdot \frac{1}{64} = .3125.$$

Example: A die is rolled four times. What is the probability that we obtain exactly one 6? We treat this as Bernoulli trials with success = "rolling a 6" and failure = "rolling some number other than a 6." Then p = 1/6, and the probability of exactly one success in four trials is:

$$b(4, 1/6, 1) = \binom{4}{1}\left(\frac{1}{6}\right)^1\left(\frac{5}{6}\right)^3 = .386.$$

To compute binomial probabilities using the computer, multiply the function choose(n, k) by $p^k q^{n-k}$. The programme *BinomialProbabilities* prints out the binomial probabilities $b(n, p, k)$ for k between k_{min} and k_{max}, and the sum of these probabilities. We have run this programme for $n = 100$, $p = 1/2$, $k_{min} = 45$, and $k_{max} = 55$; the output is shown in following table. Note that the individual probabilities are quite small. The probability of exactly 50 heads in 100 tosses of a coin is about .08. Our intuition tells us that this is the most likely outcome, which is correct; but, all the same, it is not a very likely outcome.

Table: Binomial probabilities for $n = 100$, $p = 1/2$.

k	$b(n, p, k)$
45	.0485
46	.0580
47	.0666
48	.0735
49	.0780
50	.0796
51	.0780
52	.0735

53	.0666
54	.0580
55	.0485

Binomial Distributions

Definition: Let *n* be a positive integer, and let p be a real number between 0 and 1. Let B be the random variable which counts the number of successes in a Bernoulli trials process with parameters *n* and *p*. Then, the distribution b(n, p, k) of *B* is called the binomial distribution.

We can get a better idea about the binomial distribution by graphing this distribution for different values of *n* and p. The plots in this figure were generated using the programme *BinomialPlot*.

We have run this programme for $p = .5$ and $p = .3$. Note that even for $p = .3$ the graphs are quite symmetric. We also note that the highest probability occurs around the value *np*, but that these highest probabilities get smaller as *n* increases. Np is the mean or expected value of the binomial distribution $b(n, p, k)$.

The following example gives a nice way to see the binomial distribution, when $p = 1/2$.

Example: A Galton board is a board in which a large number of BB-shots are dropped from a chute at the top of the board and deflected off a number of pins on their way down to the bottom of the board. The final position of each slot is the result of a number of random deflections either to the left or the right. We have written a programme *GaltonBoard* to simulate this experiment.

We have run the programme for the case of 20 rows of pins and 10,000 shots being dropped. We show the result of this simulation in the figure below.

Note that if we write 0 every time the shot is deflected to the left, and 1 every time it is deflected to the right, then the path of the shot can be described by a sequence of 0's and 1's of length *n*, just as for the *n*-fold coin toss.

The distribution shown in the figure below is an example of an empirical distribution, in the sense that it comes about by means of a sequence of experiments.

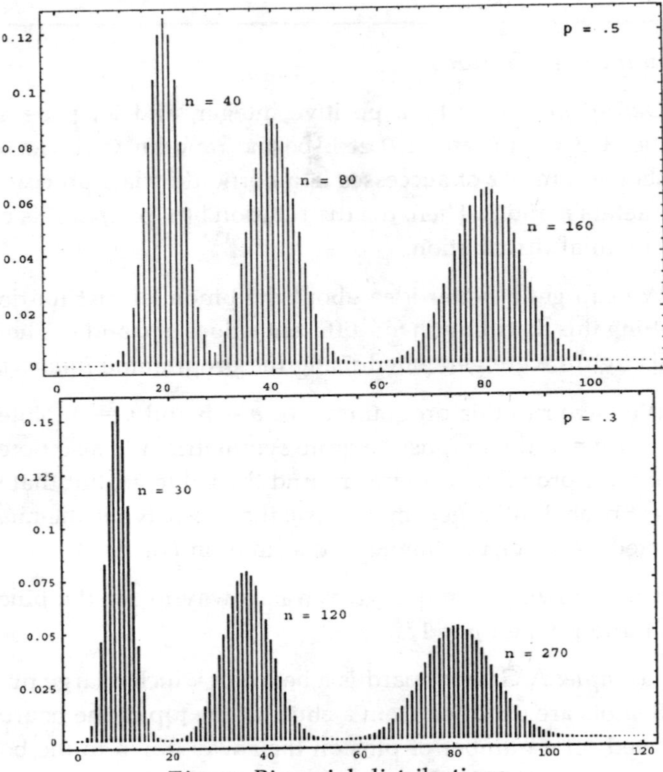

Figure: Binomial distributions.

Figure: Simulation of the Galton board.

As expected, this empirical distribution resembles the corresponding binomial distribution with parameters $n = 20$ and $p = 1/2$.

Hypothesis Testing

Example (B): Suppose that ordinary aspirin has been found effective against headaches 60 per cent of the time, and that a drug company claims that its new aspirin with a special headache additive is more effective. We can test this claim as follows: we call their claim the alternate hypothesis, and its negation, that the additive has no appreciable effect, the null hypothesis. Thus, the null hypothesis is that $p = .6$, and the alternate hypothesis is that $p > .6$, where p is the probability that the new aspirin is effective.

We give the aspirin to n people to take when they have a headache. We want to find a number m, called the critical value for our experiment, such that we reject the null hypothesis if at least m people are cured, and otherwise we accept it. How should we determine this critical value?

First note that we can make two kinds of errors. The first, often called a type 1 error in statistics, is to reject the null hypothesis when in fact it is true. The second, called a type 2 error, is to accept the null hypothesis when it is false. To determine the probability of both these types of errors, we introduce a function $\alpha(p)$, defined to be the probability that we reject the null hypothesis, where this probability is calculated under the assumption that the null hypothesis is true. In the present case, we have

$$\alpha(p) = \sum_{m \le k \le n} b(n, p, k).$$

Note that $\alpha(.6)$ is the probability of a type 1 error, since this is the probability of a high number of successes for an ineffective additive. So for a given n we want to choose m so as to make $\alpha(.6)$ quite small, to reduce the likelihood of a type 1 error. But as m increases above the most probable value $np = .6n$, $\alpha(.6)$, being the upper tail of a binomial distribution, approaches 0. Thus, increasing m makes a type 1 error less likely.

Now suppose that the additive really is effective, so that p is appreciably greater than .6; say $p = .8$. (This alternative value of p is chosen arbitrarily; the following calculations depend on this choice.) Then choosing m well below $np = .8n$ will increase $\alpha(.8)$, since now $\beta(.8)$ is all but the lower tail of a binomial distribution. Indeed, if we put $\beta(.8) = 1 - \alpha(.8)$, then $\beta(.8)$ gives us the probability of a type 2 error, and so decreasing m makes a type 2 error less likely.

The manufacturer would like to guard against a type 2 error, since if such an error is made, then the test does not show that the new drug is better, when in fact it is. If the alternative value of p is chosen closer to the value of p given in the null hypothesis (in this case $p = .6$), then for a given test population, the value of β will increase. So, if the manufacturer's statistician chooses an alternative value for p which is close to the value in the null hypothesis, then it will be an expensive proposition (i.e., the test population will have to be large) to reject the null hypothesis with a small value of β.

What we hope to do then, for a given test population n, is to choose a value of m, if possible, which makes both these probabilities small. If we make a type 1 error, we end up buying a lot of essentially ordinary aspirin at an inflated price; a type 2 error means we miss a bargain on a superior medication. Let us say that we want our critical number m to make each of these undesirable cases less than 5 per cent probable.

We write a programme *PowerCurve* to plot, for $n = 100$ and selected values of m, the function $\alpha(p)$, for p ranging from .4 to 1. The result is shown in the following figure. We include in our graph a box (in dotted lines) from .6 to .8, with bottom and top at heights .05 and .95. Then a value for m satisfies our requirements if and only if the graph of α enters the box from the bottom, and leaves from the top (why?—which is the type 1 and which is the type 2 criterion?). As m increases, the graph of α moves to the right. A few experiments have shown us that $m = 69$ is the smallest value for m that thwarts a type 1 error, while $m = 73$ is the largest which thwarts a type 2. So we may choose our critical value between 69 and 73. If we're more intent on avoiding a type 1 error we favour 73, and similarly we favour 69 if we regard a type 2 error as worse. Of course, the drug company may not be happy with having as much as a 5 per cent chance of an error. They might insist on having a 1 per cent chance of an error. For this, we would have to increase the number n of trials.

Binomial Expansion

We next remind the reader of an application of the binomial coefficients to algebra. This is the binomial expansion, from which we get the term binomial coefficient.

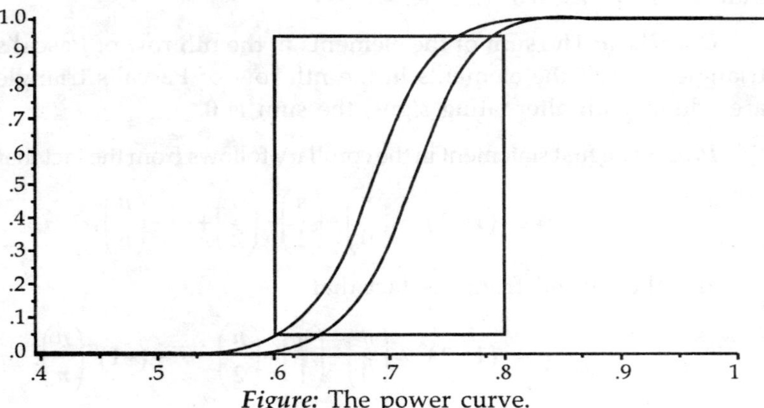

Figure: The power curve.

Theorem (g): (Binomial Theorem) The quantity $(a + b)^n$ can be expressed in the form

$$(a + b)^n = \sum_{j=0}^{n} \binom{n}{j} a^j b^{n-j}$$

Proof: To see that this expansion is correct, write

$$(a + b)^n = (a + b)(a + b). . . (a + b).$$

When we multiply this out we will have a sum of terms each of which results from a choice of an a or b for each of n factors. When we choose j a's and $(n - j)$ b's, we obtain a term of the form $a^j b^{n-j}$. To determine such a term, we have to specify j of the n terms in the products from which we choose the a. This can be done in $\binom{n}{j}$ ways. Thus, collecting these terms in the sum contributes a term $\binom{n}{j}a^j b^{n-j}$.

For example, we have

$$(a + b)^0 = 1$$

$$(a + b)^1 = a + b$$

$$(a + b)^2 = a^2 + 2ab + b^2$$

$$(a + b)^3 = a^3 + 3a^2b + 3ab^2 + b^3.$$

We see here that the coefficients of successive powers do indeed yield Pascal's triangle.

Corollary: The sum of the elements in the nth row of Pascal's triangle is 2^n. If the elements in the nth row of Pascal's triangle are added with alternating signs, the sum is 0.

Proof: The first statement in the corollary follows from the fact that

$$2^n = (1+1)^n = \binom{n}{0} + \binom{n}{1} + \binom{n}{2} + \cdots + \binom{n}{n},$$

and the second from the fact that

$$0 = (1-1)^n = \binom{n}{0} - \binom{n}{1} + \binom{n}{2} - \cdots + (-1)^n\binom{n}{n}.$$

The first statement of the corollary tells us that the number of subsets of a set of n elements is 2^n. We shall use the second statement in our next application of the binomial theorem.

We have seen that, when A and B are any two events,

$$P(A \cup B) = P(A) + P(B) - P(A \cap B).$$

We now extend this theorem to a more general version, which will enable us to find the probability that at least one of a number of events occurs.

Inclusion-Exclusion Principle

Theorem (h): Let P be a probability measure on a sample space Ω, and let $\{A_1, A_2, \ldots, A_n\}$ be a finite set of events. Then

$$P(A_1 \cup A_2 \cup \ldots A_n) = \sum_{i=1}^{n} P(A_i) - \sum_{1 \le i < j \le n} P(A_i \cap A_j)$$

$$+ \sum_{1 \le i < j < k \le n} P(A_i \cap A_j \cap A_k) - \cdots \text{(m)} \tag{5.3}$$

That is, to find the probability that at least one of n events A_i occurs, first add the probability of each event, then subtract the probabilities of all possible two-way intersections, add the probability of all three-way intersections, and so forth.

Proof: If the outcome w occurs in at least one of the events A_i, its probability is added exactly once by the left side of equation (5.3). We must show that it is added exactly once by the right side of equation (5.3). Assume that w is in exactly k of the sets. Then its probability is added k times in the first term, subtracted $\binom{k}{2}$ times in the second, added $\binom{k}{3}$ times in the third term, and so forth. Thus, the total number of times that it is added is

$$\binom{k}{1} - \binom{k}{2} + \binom{k}{3} - \cdots (-1)^{k-1} \binom{k}{k}.$$

But

$$0 = (1-1)^k = \sum_{j=0}^{k} \binom{k}{j} (-1)^j = \binom{k}{0} - \sum_{j=1}^{k} \binom{k}{j} (-1)^{j-1}.$$

Hence,

$$1 = \binom{k}{0} = \sum_{j=1}^{k} \binom{k}{j} (-1)^{j-1}.$$

If the outcome w is not in any of the events A_i, then it is not counted on either side of the equation.

Hat Check Problem

Example: We return to the hat check problem, that is, the problem of finding the probability that a random permutation contains at least one fixed point. Recall that a permutation is a one-to-one map of a set $A = \{a_1, a_2, \ldots, a_n\}$ onto itself. Let A_i be the event that the ith element a_i remains fixed under this map. If we require that a_i is fixed, then the map of the remaining $n - 1$ elements provides an arbitrary permutation of $(n - 1)$ objects. Since there are $(n - 1)!$ such permutations, $P(A_i) = (n - 1)!/n! = 1/n$. Since there are n choices for a_i, the first term of equation (5.3) is 1. In the same way, to have a particular pair (a_i, a_j) fixed, we can choose any permutation of the remaining $n - 2$ elements; there are $(n - 2)!$ such choices and thus

$$P\left(A_i \cap A_j\right) = \frac{(n-2)!}{n!} = \frac{1}{n(n-1)}.$$

The number of terms of this form in the right side of equation m is

$$\binom{n}{2} = \frac{n(n-1)}{2!}.$$

Hence, the second term of equation m is

$$-\frac{n(n-1)}{2!} \cdot \frac{1}{n(n-1)} = -\frac{1}{2!}$$

Similarly, for any specific three events A_i, A_j, A_k,

$$P\left(A_i \cap A_j \cap A_k\right) = \frac{(n-3)!}{n!} = \frac{1}{n(n-1)(n-2)},$$

and the number of such terms is

$$\binom{n}{3} = \frac{n(n-1)(n-2)}{3!},$$

making the third term of equation (5.3) equal to $1/3!$. Continuing in this way, we obtain

$$P(\text{at least one fixed point}) = 1 - \frac{1}{2!} + \frac{1}{3!} - \cdots (-1)^{n-1} \frac{1}{n!}$$

and

$$P(\text{no fixed point}) = \frac{1}{2!} - \frac{1}{3!} + \cdots (-1)^{n} \frac{1}{n!}$$

Table: Hat check problem.

n	*Probability that no one gets his own hat back*
3	.333333
4	.375
5	.366667
6	.368056
7	.367857
8	.367882
9	.367879
10	.367879

From calculus, we learn that

$$e^x = 1 + x + \frac{1}{2!}x^2 + \frac{1}{3!}x^3 + \cdots + \frac{1}{n!}x^n + \cdots.$$

Thus, if $x = -1$, we have

$$e^{-1} = \frac{1}{2!} - \frac{1}{3!} + \cdots + \frac{(-1)^n}{n!} + \cdots$$

$$= .3678794.$$

Therefore, the probability that there is no fixed point, i.e., that none of the n people gets his own hat back, is equal to the sum of the first n terms in the expression for e^{-1}. This series converges very fast. Calculating the partial sums for $n = 3$ to 10 gives the data in the previous table.

After $n = 9$ the probabilities are essentially the same to six significant figures. Interestingly, the probability of no fixed point alternately increases and decreases as n increases. Finally, we note that our exact results are in good agreement with our simulations.

Choosing a Sample Space

We now have some of the tools needed to accurately describe sample spaces and to assign probability functions to those sample spaces. Nevertheless, in some cases, the description and assignment process is somewhat arbitrary. Of course, it is to be hoped that the description of the sample space and the subsequent assignment of a probability function will yield a model which accurately predicts what would happen if the experiment were actually carried out. As the following examples show, there are situations in which "reasonable" descriptions of the sample space do not produce a model which fits the data.

In Feller's book, a pair of models is given which describe arrangements of certain kinds of elementary particles, such as photons and protons. It turns out that experiments have shown that certain types of elementary particles exhibit behaviour which is accurately described by one model, called "Bose-Einstein statistics," while other types of elementary particles can be modelled using "Fermi-Dirac statistics." Feller says:

> We have here an instructive example of the impossibility of selecting or justifying probability models by a priori arguments. In fact, no pure reasoning could tell that photons and protons would not obey the same probability laws.

We now give some examples of this description and assignment process.

Example: In the quantum mechanical model of the helium atom, various parameters can be used to classify the energy states of the atom. In the triplet spin state ($S = 1$) with orbital angular momentum 1 ($L = 1$), there are three possibilities, 0, 1, or 2, for the total angular momentum (J). (It is not assumed that the reader knows what any of this means; in fact, the example is more illustrative if the reader does not know anything about quantum mechanics.) We would like to assign probabilities to the three possibilities for J. The reader is undoubtedly resisting the idea of assigning the probability of 1/3 to each of these outcomes. She should now ask herself why she is resisting this assignment. The answer is probably because she does not have any "intuition" (i.e., experience) about the way in which helium atoms behave. In fact, in this example, the probabilities 1/9, 3/9, and 5/9 are assigned by the theory. The theory gives these assignments because these frequencies were observed *in experiments* and further parameters were developed in the theory to allow these frequencies to be predicted.

Example: Suppose two pennies are flipped once each. There are several "reasonable" ways to describe the sample space. One way is to count the number of heads in the outcome; in this case, the sample space can be written {0, 1, 2}. Another description of the sample space is the set of all ordered pairs of *H*'s and *T*'s, i.e.,

$$\{(H, H), (H, T), (T, H), (T, T)\}.$$

Both of these descriptions are accurate ones, but it is easy to see that (at most) one of these, if assigned a constant probability function, can claim to accurately model reality. In this case, as opposed to the preceding example, the reader will probably say that the second description, with each outcome being assigned a probability of 1/4, is the "right" description. This conviction is due to experience; there is no proof that this is the way reality works.

The reader is also referred to Exercise 26 for another example of this process.

Historical Remarks

The binomial coefficients have a long and colourful history leading up to Pascal's Treatise on the Arithmetical Triangle, where Pascal developed many important properties of these numbers. This history is set forth in the book Pascal's Arithmetical Triangle by A. W. F. Edwards. Pascal wrote his triangle in the form shown in the following table.

Table: Pascal's triangle.

1	1	1	1	1	1	1	1	1	1
1	2	3	4	5	6	7	8	9	
1	3	6	10	15	21	28	36		
1	4	10	20	35	56	84			
1	5	15	35	70	126				
1	6	21	56	126					
1	7	28	84						
1	8	36							
1	9								
1									

Table: Figurate numbers.

natural numbers	1	2	3	4	5	6	7	8	9
triangular numbers	1	3	6	10	15	21	28	36	45
tetrahedral numbers	1	4	10	20	35	56	84	120	165

Edwards traces three different ways that the binomial coefficients arose. He refers to these as the figurate numbers, the combinatorial numbers, and the binomial numbers. They are all names for the same thing (which we have called binomial coefficients) but that they are all the same was not appreciated until the sixteenth century.

The figurate numbers date back to the Pythagorean interest in number patterns around 540 BC. The Pythagoreans considered, for example, triangular patterns shown in the following figure. The sequence of numbers

$$1, 3, 6, 10, \ldots$$

obtained as the number of points in each triangle are called triangular numbers. From the triangles, it is clear that the nth triangular number is simply the sum of the first n integers. The tetrahedral numbers are the sums of the triangular numbers and were obtained by the Greek mathematicians Theon and Nicomachus at the beginning of the second century BC. The tetrahedral number 10, for example, has the geometric representation shown in the figure below. The first three types of figurate numbers can be represented in tabular form as shown in the previous table.

These numbers provide the first four rows of Pascal's triangle, but the table was not to be completed in the West until the sixteenth century.

In the East, Hindu mathematicians began to encounter the binomial coefficients in combinatorial problems. Bhaskara in his Lilavati of 1150 gave a rule to find the

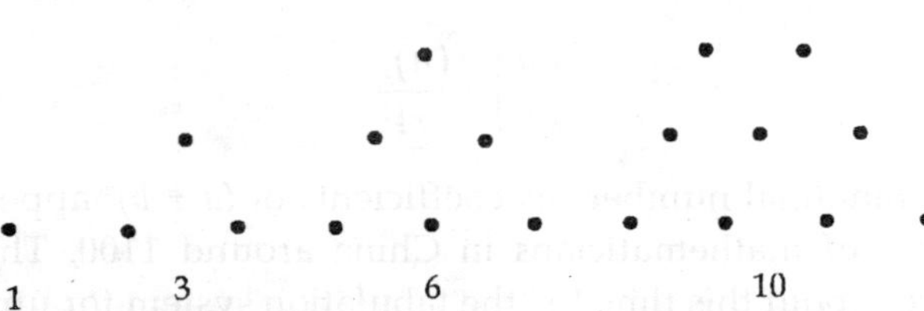

Figure: Pythagorean triangular patterns.

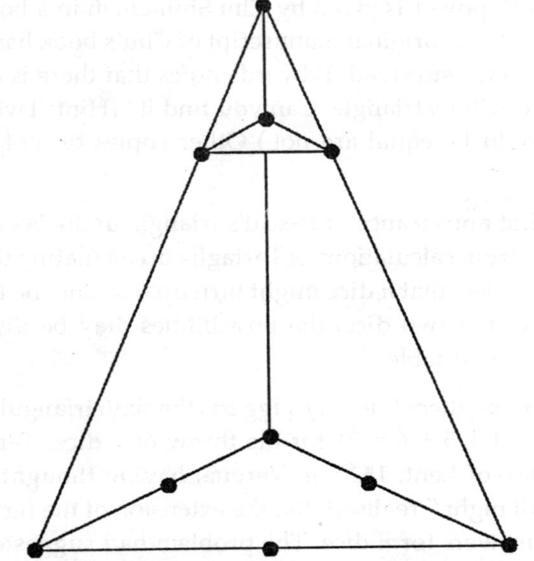

Figure: Geometric representation of the tetrahedral number 10.

Table: Outcomes for the roll of two dice.

11				
1222				
1323	33			
1424	34	44		
1525	35	45	55	
1626	36	46	56	66

number of medicinal preparations using 1, 2, 3, 4, 5, or 6 possible ingredients. His rule is equivalent to our formula

$$\binom{n}{r} = \frac{(n)_r}{r!}$$

The binomial numbers as coefficients of $(a + b)^n$ appeared in the works of mathematicians in China around 1100. There are references about this time to "the tabulation system for unlocking binomial coefficients." The triangle to provide the coefficients up to the eighth power is given by Chu Shih-chieh in a book written around 1303. The original manuscript of Chu's book has been lost, but copies have survived. Edwards notes that there is an error in this copy of Chu's triangle. Can you find it? (Hint: Two numbers which should be equal are not.) Other copies do not show this error.

The first appearance of Pascal's triangle in the West seems to have come from calculations of Tartaglia in calculating the number of possible ways that n dice might turn up. For one die the answer is clearly 6. For two dice, the possibilities may be displayed as shown in above table.

Displaying them this way suggests the sixth triangular number $1 + 2 + 3 + 4 + 5 + 6 = 21$ for the throw of 2 dice. Tartaglia "on the first day of Lent, 1523, in Verona, having thought about the problem all night," realised that the extension of the figurate table gave the answers for n dice. The problem had suggested itself to Tartaglia from watching people casting their own horoscopes by means of a Book of Fortune, selecting verses by a process which included noting the numbers on the faces of three dice. The 56 ways that three dice can fall were set out on each page. The way the numbers were written in the book did not suggest the connection with figurate numbers, but a method of enumeration similar to the one we used for 2 dice does. Tartaglia's table was not published until 1556.

A table for the binomial coefficients was published in 1554 by the German mathematician Stifel. Pascal's triangle appears also in Cardano's Opus novum of 1570.

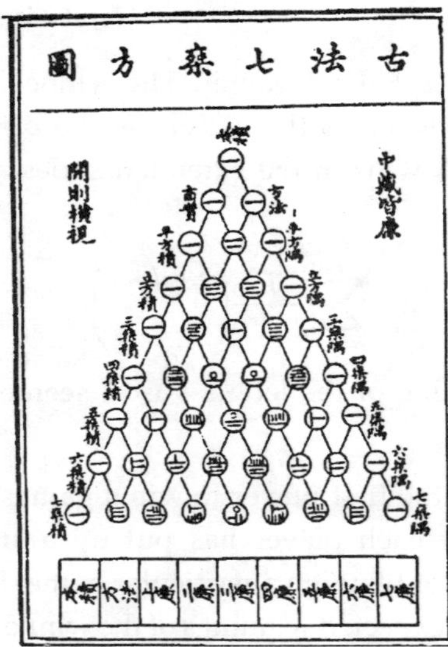

Figure: Chu Shih-chieh's triangle.

Cardano was interested in the problem of finding the number of ways to choose *r* objects out of *n*. Thus, by the time of Pascal's work, his triangle had appeared as a result of looking at the figurate numbers, the combinatorial numbers, and the binomial numbers, and the fact that all three were the same was presumably pretty well understood.

Pascal's interest in the binomial numbers came from his letters with Fermat concerning a problem known as the problem of points. The reader will recall that this problem can be described as follows: Two players A and B are playing a sequence of games and the first player to win *n* games wins the match. It is desired to find the probability that A wins the match at a time when A has won a games and B has won *b* games.

Pascal solved the problem by backward induction, much the way we would do today in writing a computer programme for its solution. He referred to the combinatorial method of Fermat which proceeds as follows: If A needs c games and B needs d games to win, we require that the players continue to play until they have played $c + d - 1$ games. The winner in this extended series will be the same as the winner in the original series. The probability that A wins in the extended series and hence in the original series is

$$\sum_{r=c}^{c+d-1} \frac{1}{2^{c+d-1}} \binom{c+d-1}{r}$$

Even at the time of the letters Pascal seemed to understand this formula.

Suppose that the first player to win n games wins the match, and suppose that each player has put up a stake of x. Pascal studied the value of winning a particular game. By this, he meant the increase in the expected winnings of the winner of the particular game under consideration. He showed that the value of the first game is

$$\frac{1 \cdot 3 \cdot 5 \cdot \ldots \cdot (2n-1)}{2 \cdot 4 \cdot 6 \cdot \ldots \cdot (2n)} x .$$

His proof of this seems to use Fermat's formula and the fact that the above ratio of products of odd to products of even numbers is equal to the probability of exactly n heads in $2n$ tosses of a coin.

Pascal presented Fermat with the table shown in the following table. He states:

> You will see as always, that the value of the first game is equal to that of the second which is easily shown by combinations. You will see, in the same way, that the numbers in the first line are always increasing; so also are those in the second; and those in the third. But those in the fourth line are decreasing, and those in the fifth, etc. This seems odd.

The student can pursue this question further using the computer and Pascal's backward iteration method for computing the expected payoff at any point in the series.

Table: Pascal's solution for the problem of points.

From my opponent's 256 positions I get, for the	if each one staken 256 in 6 games	5 games	4 games	3 games	2 games	1 games
1st game	63	70	80	96	128	256
2nd game	63	70	80	96	128	
3rd game	56	60	64	64		
4th game	42	40	32			
5th game	24	16				
6th game	8					

In his treatise, Pascal gave a formal proof of Fermat's combinatorial formula as well as proofs of many other basic properties of binomial numbers. Many of his proofs involved induction and represent some of the first proofs by this method. His book brought together all the different aspects of the numbers in the Pascal triangle as known in 1654, and, as Edwards states, "That the Arithmetical Triangle should bear Pascal's name cannot be disputed."

The first serious study of the binomial distribution was undertaken by James Bernoulli in his Ars Conjectandi published in 1713.

Exercises

1. Compute the following:

   (a) $\binom{6}{3}$

   (b) $b(5, .2, 4)$

   (c) $\binom{7}{2}$

   (d) $\binom{26}{26}$

   (e) $b(4, .2, 3)$

   (f) $\binom{6}{2}$

   (g) $\binom{10}{9}$

   (h) $b(8, .3, 5)$.

2. In how many ways can we choose five people from a group of ten to form a committee?

3. How many seven-element subsets are there in a set of nine elements?

4. Using the relation equation y, write a programme to compute Pascal's triangle, putting the results in a matrix. Have your programme print the triangle for $n = 10$.

5. Use the programme BinomialProbabilities to find the probability that, in 100 tosses of a fair coin, the number of heads that turns up lies between 35 and 65, between 40 and 60, and between 45 and 55.

6. Charles claims that he can distinguish between beer and ale 75 per cent of the time. Ruth bets that he cannot and, in fact, just guesses. To settle this, a bet is made: Charles is to be given ten small glasses, each having been filled with beer or ale, chosen by tossing a fair coin. He wins the bet if he gets seven or more correct. Find the probability that Charles wins if he has the ability that he claims. Find the probability that Ruth wins if Charles is guessing.

7. Show that

$$b(n,p,j) = \frac{p}{q}\left(\frac{n-j+1}{j}\right)b(n,p,j-1),$$

for $j \geq 1$. Use this fact to determine the value or values of j which give $b(n, p, j)$ its greatest value. *Hint*: Consider the successive ratios as j increases.

8. A die is rolled 30 times. What is the probability that a 6 turns up exactly 5 times? What is the most probable number of times that a 6 will turn up?

9. Find integers n and r such that the following equation is true:

$$\binom{13}{5} + 2\binom{13}{6} + \binom{13}{7} = \binom{n}{r}$$

10. In a ten-question true-false exam, find the probability that a student gets a grade of 70 per cent or better by guessing.

Answer the same question if the test has 30 questions, and if the test has 50 questions.

11. A restaurant offers apple and blueberry pies and stocks an equal number of each kind of pie. Each day, ten customers request pie. They choose, with equal probabilities, one of the two kinds of pie. How many pieces of each kind of pie should the owner provide so that the probability is about .95 that each customer gets the pie of his or her own choice?

12. A poker hand is a set of 5 cards randomly chosen from a deck of 52 cards.

    Find the probability of a

    (a) royal flush (ten, jack, Queen, King, ace in a single suit).

    (b) straight flush (five in a sequence in a single suit, but not a royal flush).

    (c) four of a kind (four cards of the same face value).

    (d) full house (one pair and one triple, each of the same face value).

    (e) flush (five cards in a single suit but not a straight or royal flush).

    (f) straight (five cards in a sequence, not all the same suit). (Note that in straights, an ace counts high or low.)

13. If a set has $2n$ elements, show that it has more subsets with n elements than with any other number of elements.

14. Let $b(2n, .5, n)$ be the probability that in 2n tosses of a fair coin exactly n heads turn up. Using Stirling's formula (Theorem c), show that $b(2n, .5, n) 1/\sqrt{\pi n}$. Use the programme *BinomialProbabilities* to compare this with the exact value for $n = 10$ to 25.

15. A baseball player, Smith, has a batting average of .300 and in a typical game comes to bat three times. Assume that Smith's hits in a game can be considered to be a Bernoulli trials process with probability .3 for success. Find the probability that Smith gets 0, 1, 2, and 3 hits.

16. The Siwash University football team plays eight games in a season, winning three, losing three, and ending two in a tie. Show that the number of ways that this can happen is

$$\binom{8}{3}\binom{5}{3} = \frac{8!}{3!3!2!}.$$

17. Using the technique of Exercise 16, show that the number of ways that one can put n different objects into three boxes with a in the first, b in the second, and c in the third is $n!/(a!\ b!\ c!)$.

18. Baumgartner, Prosser, and Crowell are grading a calculus exam. There is a true-false question with ten parts. Baumgartner notices that one student has only two out of the ten correct and remarks, "The student was not even bright enough to have flipped a coin to determine his answers." "Not so clear," says Prosser. "With 340 students I bet that if they all flipped coins to determine their answers there would be at least one exam with two or fewer answers correct." Crowell says, "I'm with Prosser. In fact, I bet that we should expect at least one exam in which no answer is correct if everyone is just guessing." Who is right in all of this?

19. A gin hand consists of 10 cards from a deck of 52 cards. Find the probability that a gin hand has:
    (a) all 10 cards of the same suit.
    (b) exactly 4 cards in one suit and 3 in two other suits.
    (c) a 4, 3, 2, 1, distribution of suits.

20. A six-card hand is dealt from an ordinary deck of cards. Find the probability that:
    (a) All six cards are hearts.
    (b) There are three aces, two Kings, and one Queen.
    (c) There are three cards of one suit and three of another suit.

21. A lady wishes to colour her fingernails on one hand using at most two of the colours red, yellow, and blue. How many ways can she do this?

22. How many ways can six indistinguishable letters be put in three mail boxes? *Hint*: One representation of this is given by a sequence |LL|L|LLL| where the |'s represent the partitions for the boxes and the L's the letters. Any possible way can be so described. Note that we need two bars at the ends and the remaining two bars and the six L's can be put in any order.

23. Using the method for the hint in Exercise 22, show that r indistinguishable objects can be put in n boxes in different ways.

$$\binom{n+r-1}{n-1} = \binom{n+r-1}{r}$$

24. A travel bureau estimates that when 20 tourists go to a resort with ten hotels they distribute themselves as if the bureau were putting 20 indistinguishable objects into ten distinguishable boxes. Assuming this model is correct, find the probability that no hotel is left vacant when the first group of 20 tourists arrives.

25. An elevator takes on six passengers and stops at ten floors. We can assign two different equiprobable measures for the ways that the passengers are discharged: (a) we consider the passengers to be distinguishable or (b) we consider them to be indistinguishable. For each case, calculate the probability that all the passengers get off at different floors.

26. You are playing heads or tails with Prosser but you suspect that his coin is unfair. Von Neumann suggested that you proceed as follows: Toss Prosser's coin twice. If the outcome is HT call the result win. If it is TH call the result lose. If it is TT or HH ignore the outcome and toss Prosser's coin twice again. Keep going until you get either an HT or a TH and call the result win or lose in a single play. Repeat this

procedure for each play. Assume that Prosser's coin turns up heads with probability p.

(a) Find the probability of HT, TH, HH, TT with two tosses of Prosser's coin;

(b) Using part (a), show that the probability of a win on any one play is 1/2, no matter what p is.

27. John claims that he has extrasensory powers and can tell which of two symbols is on a card turned face down. To test his ability he is asked to do this for a sequence of trials. Let the null hypothesis be that he is just guessing, so that the probability is 1/2 of his getting it right each time, and let the alternative hypothesis be that he can name the symbol correctly more than half the time. Devise a test with the property that the probability of a type 1 error is less than .05 and the probability of a type 2 error is less than .05 if John can name the symbol correctly 75 per cent of the time.

28. Assume the alternative hypothesis is that p = .8 and that it is desired to have the probability of each type of error less than .01. Use the programme *PowerCurve* to determine values of n and m that will achieve this. Choose n as small as possible.

29. A drug is assumed to be effective with an unknown probability p. To estimate p the drug is given to n patients. It is found to be effective for m patients. The method of maximum likelihood for estimating p states that we should choose the value for p that gives the highest probability of getting what we got on the experiment. Assuming that the experiment can be considered as a Bernoulli trials process with probability p for success, show that the maximum likelihood estimate for p is the proportion m/n of successes.

30. Recall that in the World Series the first team to win four games wins the series. The series can go at most seven games. Assume that the Red Sox and the Mets are playing the series. Assume that the Mets win each game with probability p.

Fermat observed that even though the series might not go seven games, the probability that the Mets win the series is the same as the probability that they win four or more game in a series that was forced to go seven games no matter who wins the individual games.

(a) Using the programme *Power Curve* the probability that the Mets win the series for the cases $p = .5$, $p = .6$, $p = .7$.

(b) Assume that the Mets have probability .6 of winning each game. Use the programme *PowerCurve* to find a value of n so that, if the series goes to the first team to win more than half the games, the Mets will have a 95 per cent chance of winning the series. Choose n as small as possible.

31. Each of the four engines on an airplane functions correctly on a given flight with probability .99, and the engines function independently of each other. Assume that the plane can make a safe landing if at least two of its engines are.functioning correctly. What is the probability that the engines will allow for a safe landing?

32. A small boy is lost coming down Mount Washington. The leader of the search team estimates that there is a probability p that he came down on the east side and a probability $1 - p$ that he came down on the west side. He has n people in his search team who will search independently and, if the boy is on the side being searched, each member will find the boy with probability u. Determine how he should divide the n people into two groups to search the two sides of the mountain so that he will have the highest probability of finding the boy. How does this depend on u?

33. $2n$ balls are chosen at random from a total of $2n$ red balls and $2n$ blue balls. Find a combinatorial expression for the probability that the chosen balls are equally divided in colour. Use Stirling's formula to estimate this probability.

    Using *BinomialProbabilities*, compare the exact value with Stirling's approximation for $n = 20$.

34. Assume that every time you buy a box of Wheaties, you receive one of the pictures of the n players on the New York Yankees. Over a period of time, you buy $m \geq n$ boxes of Wheaties.

    (a) Use Theorem (h) to show that the probability that you get all n pictures

    $$1-\binom{n}{1}\left(\frac{n-1}{n}\right)^m+\binom{n}{2}\left(\frac{n-2}{n}\right)^m-\cdots+(-1)^{n-1}\binom{n}{n-1}\left(\frac{1}{n}\right)^m.$$

    Hint: Let E_k be the event that you do not get the kth player's picture.

    (b) Write a computer programme to compute this probability. Use this programme to find, for given n, the smallest value of m which will give probability $\geq .5$ of getting all n pictures. Consider $n = 50$, 100, and 150 and show that $m = n\log n + n\log 2$ is a good estimate for the number of boxes needed.

35. Prove the following binomial identity

    $$\binom{2n}{n}=\sum_{j=0}^{n}\binom{n}{j}^2.$$

    Hint: Consider an urn with n red balls and n blue balls inside. Show that each side of the equation equals the number of ways to choose n balls from the urn.

36. Let j and n be positive integers, with $j \leq n$. An experiment consists of choosing, at random, a j-tuple of positive integers whose sum is at most n.

    (a) Find the size of the sample space. *Hint*: Consider n indistinguishable balls placed in a row. Place j markers between consecutive pairs of balls, with no two markers between the same pair of balls. (We also allow one of the n markers to be placed at the end of the row of balls). Show that there is a 1-1 correspondence between the set of possible positions for the markers and the set of j-tuples whose size we are trying to count.

(b) Find the probability that the j -tuple selected contains at least one 1.

37. Let $n \pmod m$ denote the remainder when the integer n is divided by the integer m. Write a computer programme to compute the number $\binom{n}{j} \pmod m$, where $\binom{n}{j}$ is binomial coefficient and m is an integer. You can do this by using the recursion relations for generating binomial coefficients, doing all the arithmetic using the basic function mod(n, m). Try to write your programme to make as large a table as possible. Run your programme for the cases $m = 2$ to 7. Do you see any patterns? In particular, for the case m = 2 and n a power of 2, verify that all the entries in the (n – 1)st row are 1. (The corresponding binomial numbers are odd.) Use your pictures to explain why this is true.

38. Lucas proved the following general result relating to Exercise 37. If p is any prime number, then $\binom{n}{j} \pmod p$ can be found as follows: Expand n and j in base p as $n = s_0 + s_1p + s_2p^2 + \ldots + s_kp^k$ and $j = r_0 + r_1p + r_2p^2 + \ldots + r_kp^k$, respectively. (Here k is chosen large enough to represent all numbers from 0 to n in base p using k digits.) Let $s = (s_0, s_1, s_2, \ldots, s_k)$ and $r = (r_0, r_1, r_2, \ldots, r_k)$. Then

$$\binom{n}{j} \pmod p = \prod_{i=0}^{k} \binom{s_i}{r_i} \pmod p.$$

For example, if $p = 7$, $n = 12$, and $j = 9$, then

$$12 = 5 \cdot 7^0 + 1.7^1,$$

$$9 = 2 \cdot 7^0 + 1.7^1,$$

so that

$$s = (5, 1),$$

$$r = (2, 1),$$

and this result states that

$$\binom{12}{9} \pmod p = \binom{5}{2}\binom{1}{1} \pmod 7$$

Since $\binom{12}{9} = 220 \equiv 3 \pmod 7$, and $\binom{5}{2} = 10 \equiv 3 \pmod 7$, we see that the result is correct for this example.

Show that this result implies that, for $p = 2$, the $(p^k - 1)$st row of your triangle in Exercise 37 has no zeros.

39. Prove that the probability of exactly n heads in $2n$ tosses of a fair coin is given by the product of the odd numbers up to $2n - 1$ divided by the product of the even numbers up to $2n$.

40. Let n be a positive integer, and assume that j is a positive integer not exceeding $n/2$. Show that in Theorem (e), if one alternates the multiplications and divisions, then all of the intermediate values in the calculation are integers. Show also that none of these intermediate values exceed the final value.

Card Shuffling

Much of this section is based upon an article by Brad Mann, which is an exposition of an article by David Bayer and Persi Diaconis.

Riffle Shuffles

Given a deck of n cards, how many times must we shuffle it to make it "random"? Of course, the answer depends upon the method of shuffling which is used and what we mean by "random." We shall begin the study of this question by considering a standard model for the riffle shuffle.

We begin with a deck of n cards, which we will assume are labelled in increasing order with the integers from 1 to n. A riffle shuffle consists of a cut of the deck into two stacks and an interleaving of the two stacks. For example, if $n = 6$, the initial ordering is (1, 2, 3, 4, 5, 6), and a cut might occur between cards 2 and 3. This gives rise to two stacks, namely (1, 2) and (3, 4, 5, 6). These are interleaved to form a new ordering of the deck. For example, these two stacks might form the ordering (1, 3, 4, 2, 5, 6). In order to discuss such shuffles, we need to assign a probability measure to the set of all possible shuffles. There are several reasonable ways in which this can be done. We will give several different assignment strategies, and show that they are equivalent. (This does not mean that this assignment is the only reasonable one.) First, we assign the binomial probability $b(n, 1/2, k)$ to the event that the cut occurs after the kth card. Next, we assume that all possible interleavings, given a cut, are equally likely. Thus, to complete the assignment of probabilities, we need to determine the number of possible interleavings of two stacks of cards, with k and $n - k$ cards, respectively.

We begin by writing the second stack in a line, with spaces in between each pair of consecutive cards, and with spaces at the beginning and end (so there are $n - k + 1$ spaces). We choose, with replacement, k of these spaces, and place the cards from the first stack in the chosen spaces. This can be done in

$$\binom{n}{k}$$

ways. Thus, the probability of a given interleaving should be

$$\frac{1}{\binom{n}{k}}.$$

Next, we note that if the new ordering is not the identity ordering, it is the result of a unique cut-interleaving pair. If the new ordering is the identity, it is the result of any one of $n + 1$ cut-interleaving pairs.

We define a rising sequence in an ordering to be a maximal subsequence of consecutive integers in increasing order. For example, in the ordering (2, 3, 5, 1, 4, 7, 6), there are 4 rising sequences; they are (1), (2, 3, 4), (5, 6), and (7). It is easy to see that an ordering is the result of a riffle shuffle applied to the identity ordering if and only if it has no more than two rising sequences. (If the ordering has two rising sequences, then these rising sequences correspond to the two stacks induced by the cut, and if the ordering has one rising sequence, then it is the identity ordering).

Thus, the sample space of orderings obtained by applying a riffle shuffle to the identity ordering is naturally described as the set of all orderings with at most two) rising sequences.

It is now easy to assign a probability measure to this sample space. Each ordering with two rising sequences is assigned the value

$$\frac{b(n,1/2,k)}{\binom{n}{k}} = \frac{1}{2^n},$$

and the identity ordering is assigned the value

$$\frac{n+1}{2^n}.$$

There is another way to view a riffle shuffle. We can imagine starting with a deck cut into two stacks as before, with the same probabilities assignment as before, i.e., the binomial distribution. Once we have the two stacks, we take cards, one by one, off of the bottom of the two stacks, and place them onto one stack. If there are k_1 and k_2 cards, respectively, in the two stacks at some point in this process, then we make the assumption that the probabilities that the next card to be taken comes from a given stack is proportional to the current stack size. This implies that the probability that we take the next card from the first stack equals

$$\frac{k_1}{k_1 + k_2},$$

and the corresponding probability for the second stack is

$$\frac{k_2}{k_1 + k_2}$$

We shall now show that this process assigns the uniform probability to each of the possible interleavings of the two stacks.

Suppose, for example, that an interleaving came about as the result of choosing cards from the two stacks in some order. The probability that this result occurred is the product of the probabilities at each point in the process, since the choice of card at each point is assumed to be independent of the previous choices. Each factor of this product is of the form

$$\frac{k_i}{k_1 + k_2},$$

where $i = 1$ or 2, and the denominator of each factor equals the number of cards left to be chosen. Thus, the denominator of the probability is just n!. At the moment when a card is chosen from a stack that has i cards in it, the numerator of the corresponding factor in the probability is i, and the number of cards in this stack decreases by 1. Thus, the numerator is seen to be $k!(n - k)!$, since all cards in both stacks are eventually chosen. Therefore, this process assigns the probability

$$\frac{1}{\binom{n}{k}}$$

to each possible interleaving.

We now turn to the question of what happens when we riffle shuffle s times. It should be clear that if we start with the identity ordering, we obtain an ordering with at most 2^s rising sequences, since a riffle shuffle creates at most two rising sequences from every rising sequence in the starting ordering. In fact, it is not hard to see that each such ordering is the result of s riffle shuffles. The question becomes, then, in how many ways can an ordering with r rising sequences come about by applying s riffle shuffles to the identity ordering? In order to answer this question, we turn to the idea of an a-shuffle.

a-Shuffles

There are several ways to visualise an a-shuffle. One way is to imagine a creature with a hands who is given a deck of cards to riffle shuffle. The creature naturally cuts the deck into a stacks, and then riffles them together. (Imagine that!) Thus, the ordinary riffle shuffle is a 2-shuffle. As in the case of the ordinary 2-shuffle, we allow some of the stacks to have 0 cards. Another way to visualise an a-shuffle is to think about its inverse, called an a-unshuffle. This idea is described in the proof of the next theorem.

We will now show that an a-shuffle followed by a b-shuffle is equivalent to an ab-shuffle. This means, in particular, that s riffle shuffles in succession are equivalent to one 2^s-shuffle. This equivalence is made precise by the following theorem.

Theorem (i): Let a and b be two positive integers. Let $S_{a,b}$ be the set of all ordered pairs in which the first entry is an a-shuffle and the second entry is a b-shuffle. Let S_{ab} be the set of all ab-shuffles. Then there is a 1-1 correspondence between $S_{a,b}$ and S_{ab} with the following property. Suppose that (T_1, T_2) corresponds to T_3. If T_1 is applied to the identity ordering, and T_2 is applied to the resulting ordering, then the final ordering is the same as the ordering that is obtained by applying T_3 to the identity ordering.

Proof: The easiest way to describe the required correspondence is through the idea of an unshuffle. An a-unshuffle begins with a deck of n cards. One by one, cards are taken from the top of the deck and placed, with equal probability, on the bottom of any one of a stacks, where the stacks are labelled from 0 to $a - 1$. After all of the cards have been distributed, we combine the stacks to form one stack by placing stack i on top of stack $i + 1$, for $0 \le i \le a - 1$. It is easy to see that if one starts with a deck, there is exactly one way to cut the deck to obtain the a stacks generated by the a-unshuffle, and with these a stacks, there is exactly one way to interleave them to obtain the deck in the order that it was in before the unshuffle was performed. Thus, this a-unshuffle corresponds to a unique a-shuffle, and this a-shuffle is the inverse of the original a-unshuffle.

If we apply an ab-unshuffle U_3 to a deck, we obtain a set of ab stacks, which are then combined, in order to form one stack. We label these stacks with ordered pairs of integers, where the first coordinate is between 0 and $a - 1$, and the second coordinate is between 0 and $b - 1$. Then we label each card with the label of its stack. The number of possible labels is ab, as required. Using this labelling, we can describe how to find a b-unshuffle and an *a*-unshuffle, such that if these two unshuffles are applied in this order to the deck, we obtain the same set of ab stacks as were obtained by the *ab*-unshuffle.

To obtain the *b*-unshuffle U_2, we sort the deck into b stacks, with the *i*th stack containing all of the cards with second coordinate i, for $0 \le i \le b-1$. Then these stacks are combined to form one stack. The a-unshuffle U_1 proceeds in the same manner, except that the first coordinates of the labels are used. The resulting a stacks are then combined to form one stack.

The above description shows that the cards ending up on top are all those labelled (0, 0). These are followed by those labelled $(0, 1), (0, 2), \ldots, (0, b-1), (1, 0), (1, 1), \ldots, (a-1, b-1)$. Furthermore, the relative order of any pair of cards with the same labels is never altered. But this is exactly the same as an ab-unshuffle, if, at the beginning of such an unshuffle, we label each of the cards with one of the labels $(0, 0), (0, 1), \ldots, (0, b-1), (1, 0), (1, 1), \ldots, (a-1, b-1)$. This completes the proof.

In the following figure, we show the labels for a 2-unshuffle of a deck with 10 cards. There are 4 cards with the label 0 and 6 cards with the label 1, so if the 2-unshuffle is performed, the first stack will have 4 cards and the second stack will have 6 cards. When this unshuffle is performed, the deck ends up in the identity ordering.

In the figure below, we show the labels for a 4-unshuffle of the same deck (because there are four labels being used). This figure can also be regarded as an example of a pair of 2-unshuffles, as described in the proof above. The first 2-unshuffle will use the second coordinate of the labels to determine the stacks. In this case, the two stacks contain the cards whose values are:

$$\{5, 1, 6, 2, 7\} \text{ and } \{8, 9, 3, 4, 10\}.$$

After this 2-unshuffle has been performed, the deck is in the order shown in the following figure, as the reader should check.

If we wish to perform a 4-unshuffle on the deck, using the labels shown, we sort the cards lexicographically, obtaining the four stacks

$$\{1, 2\}, \{3, 4\}, \{5, 6, 7\}, \text{ and } \{8, 9, 10\}.$$

When these stacks are combined, we once again obtain the identity ordering of the deck. The point of the above theorem is that both sorting procedures always lead to the same initial ordering.

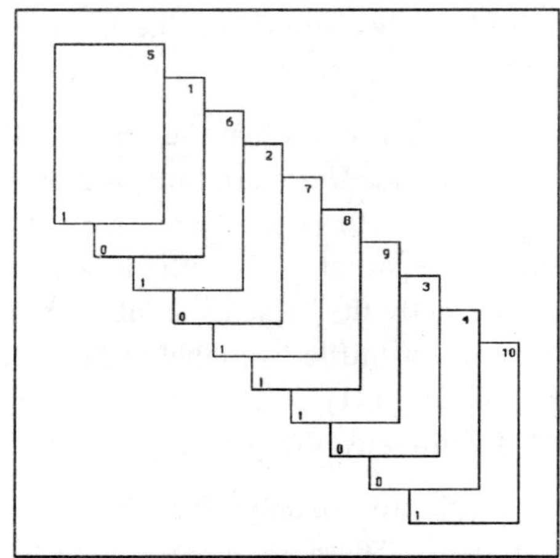

Figure: Before a 2-unshuffle.

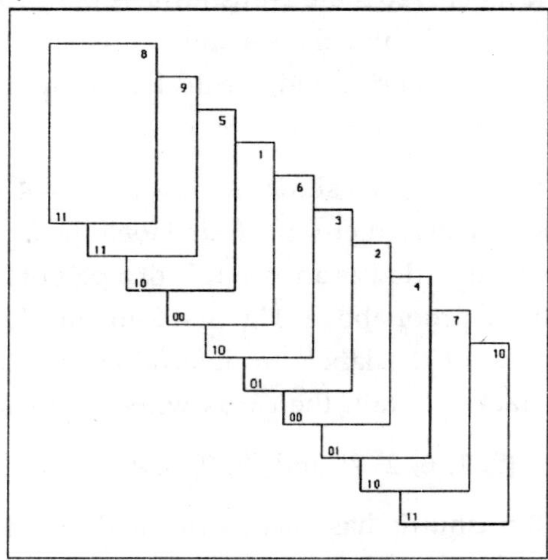

Figure: Before a 4-unshuffle.

Theorem (j): If D is any ordering that is the result of applying an a-shuffle and then *ab*-shuffle to the identity ordering, then the probability assigned to D by this pair of operations is the same as the probability assigned to D by the process of applying an *ab*-shuffle to the identity ordering.

Proof: Call the sample space of *a*-shuffles S_a. If we label the stacks by the integers from 0 to $a - 1$, then each cut-interleaving pair, i.e., shuffle, corresponds to exactly one n-digit base a integer, where the *i*th digit in the integer is the stack of which the *i*th card is a member. Thus, the number of cut-interleaving pairs is equal to the number of *n*-digit base a integers, which is a^n. Of course, not all of these pairs leads to different orderings. The number of pairs leading to a given ordering will be discussed later. For our purposes, it is enough to point out that it is the cut-interleaving pairs that determine the probability assignment.

The previous theorem shows that there is a 1-1 correspondence between $S_{a,b}$ and S_{ab}. Furthermore, corresponding elements give the same ordering when applied to the identity ordering. Given any ordering D, let m_1 be the number of elements of $S_{a,b}$ which, when applied to the identity ordering, result in D. Let m_2 be the number of elements of S_{ab} which, when applied to the identity ordering, result in D. The previous theorem implies that $m_1 = m_2$. Thus, both sets assign the probability

$$\frac{m_1}{(ab)^n}$$

to D. This completes the proof.

Connection with the Birthday Problem

There is another point that can be made concerning the labels given to the cards by the successive unshuffles. Suppose that we 2-unshuffle an *n*-card deck until the labels on the cards are all different. It is easy to see that this process produces each permutation with the same probability, i.e., this is a random process. To see this, note that if the labels become distinct on the sth 2-unshuffle, then one can think of this sequence of 2-unshuffles as one 2^s -unshuffle, in which all of the stacks determined by the unshuffle have at most one card in them (remember, the stacks correspond to the labels). If each stack has at most one card in it, then given any two cards in the deck, it is equally likely that the first card has a lower or a higher label than the second card. Thus, each possible ordering is equally likely to result from this 2^s-unshuffle.

Let T be the random variable that counts the number of 2-unshuffles until all labels are distinct. One can think of T as giving a measure of how long it takes in the unshuffling process until randomness is reached. Since shuffling and unshuffling are inverse processes, T also measures the number of shuffles necessary to achieve randomness. Suppose that we have an n-card deck, and we ask for $P(T \leq s)$. This equals $1 - P(T > s)$. But $T > s$ if and only if it is the case that not all of the labels after s 2-unshuffles are distinct. This is just the birthday problem; we are asking for the probability that at least two people have the same birthday, given that we have n people and there are 2^s possible birthdays. Using our formula, we find that

$$P(T > s) = 1 - \binom{2^s}{n} \frac{n!}{2^{sn}}. \qquad (5.4)$$

Using the average value of a random variable, and the above equation, one can calculate the average value of the random variable T. For example, if $n = 52$, then the average value of T is about 11.7. This means that, on the average, about 12 riffle shuffles are needed for the process to be considered random.

Cut-Interleaving Pairs and Orderings

As was noted in the proof of Theorem *(j)*, not all of the cut-interleaving pairs lead to different orderings. However, there is an easy formula which gives the number of such pairs that lead to a given ordering.

Theorem (k): If an ordering of length n has r rising sequences, then the number of cut-interleaving pairs under an a-shuffle of the identity ordering which lead to the ordering is

$$\binom{n+a-r}{n}.$$

Proof: To see why this is true, we need to count the number of ways in which the cut in an a-shuffle can be performed which will lead to a given ordering with r rising sequences. We can disregard the interleavings, since once a cut has been made, at most one interleaving will lead to a given ordering.

Since the given ordering has r rising sequences, $r - 1$ of the division points in the cut are determined. The remaining $a - 1 - (r - 1) = a - r$ division points can be placed anywhere.

The number of places to put these remaining division points is $n + 1$ (which is the number of spaces between the consecutive pairs of cards, including the positions at the beginning and the end of the deck). These places are chosen with repetition allowed, so the number of ways to make these choices is

$$\binom{n+a-r}{a-r} = \binom{n+a-r}{n}.$$

In particular, this means that if D is an ordering that is the result of applying an a-shuffle to the identity ordering, and if D has r rising sequences, then the probability assigned to D by this process is

$$\frac{\binom{n+a-r}{n}}{a^n}.$$

This completes the proof.

The above theorem shows that the essential information about the probability assigned to an ordering under an a-shuffle is just the number of rising sequences in the ordering. Thus, if we determine the number of orderings which contain exactly r rising sequences, for each r between 1 and n, then we will have determined the distribution function of the random variable which consists of applying a random a-shuffle to the identity ordering.

The number of orderings of $\{1, 2, \ldots, n\}$ with r rising sequences is denoted by $A(n, r)$, and is called an Eulerian number. There are many ways to calculate the values of these numbers; the following theorem gives one recursive method which follows immediately from what we already know about a-shuffles.

Theorem (l): Let a and n be positive integers. Then

$$a^n = \sum_{r=1}^{a} \binom{n+a-r}{n} A(n,r). \tag{5.5}$$

Thus,

$$A(n, a) = a^n - \sum_{r=1}^{a-1} \binom{n+a-r}{n} A(n,r).$$

In addition,

$$A(n,1) = 1$$

Proof: The second equation can be used to calculate the values of the Eulerian numbers, and follows immediately from the equation (5.5). The last equation is a consequence of the fact that the only ordering of $\{1, 2, \ldots, n\}$ with one rising sequence is the identity ordering.

Thus, it remains to prove equation (5.5). We will count the set of a-shuffles of a deck with n cards in two ways. First, we know that there are a^n such shuffles [this was noted in the proof of Theorem *(j)*]. But there are $A(n, r)$ orderings of $\{1, 2, \ldots, n\}$ with r rising sequences, and Theorem (k) states that for each such ordering, there are exactly

$$\binom{n+a-r}{n}$$

cut-interleaving pairs that lead to the ordering. Therefore, the right-hand side of equation (5.5) counts the set of a-shuffles of an n-card deck. This completes the proof.

Random Orderings and Random Processes

What do we mean by a "random" ordering? It is somewhat misleading to think about a given ordering as being random or not random. If we want to choose a random ordering from the set of all orderings of {1, 2, . . ., n}, we mean that we want every ordering to be chosen with the same probability, i.e., any ordering is as "random" as any other.

The word "random" should really be used to describe a process. We will say that a process that produces an object from a (finite) set of objects is a random process if each object in the set is produced with the same probability by the process. In the present situation, the objects are the orderings, and the process which produces these objects is the shuffling process. It is easy to see that no a-shuffle is really a random process, since if T_1 and T_2 are two orderings with a different number of rising sequences, then they are produced by an a-shuffle, applied to the identity ordering, with different probabilities.

Variation Distance

Instead of requiring that a sequence of shuffles yield a process which is random, we will define a measure that describes how far away a given process is from a random process. Let X be any process which produces an ordering of {1, 2, . . ., n}. Define $f_X(\pi)$ be the probability that X produces the ordering π. (Thus, X can be thought of as a random variable with distribution function f). Let Ω_n be the set of all orderings of {1, 2, . . ., n}. Finally, let $u(\pi) = 1/|\Omega_n|$ for all $\pi \in \Omega_n$. The function u is the distribution function of a process which produces orderings and which is random. For each ordering $\pi \in \Omega_n$, the quantity

$$|f_X(\pi) - u(\pi)|$$

is the difference between the actual and desired probabilities that X produces π. If we sum this overall orderings π and call this sum S, we see that $S = 0$ if and only if X is random, and otherwise S is positive. It is easy to show that the maximum value of S is 2, so we will multiply the sum by 1/2 so that the value falls in the interval [0, 1]. Thus, we obtain the following sum as the formula for the variation distance between the two processes:

$$\|f_X - u\| = \frac{1}{2}\sum_{\pi \in \Omega_n} |f_X(\pi) - u(\pi)|.$$

Now, we apply this idea to the case of shuffling. We let X be the process of s successive riffle shuffles applied to the identity ordering. We know that it is also possible to think of X as one 2^s -shuffle. We also know that f_X is constant on the set of all orderings with r rising sequences, where r is any positive integer. Finally, we know the value of f_X on an ordering with r rising sequences, and we know how many such orderings there are. Thus, in this specific case, we have

$$\|f_X - u\| = \frac{1}{2}\sum_{r=1} A(n,r)\left|\binom{2^s + n - r}{n} / 2^{ns} - \frac{1}{n!}\right|.$$

Since this sum has only n summands, it is easy to compute this for moderate sized values of n. For $n = 52$, we obtain the list of values given in following table.

Table: Distance to the random process.

Number of Riffle Shuffles	*Variation Distance*
1	1
2	1
3	1
4	0.9999995334
5	0.9237329294
6	0.6135495966
7	0.3340609995
8	0.1671586419
9	0.0854201934
10	0.0429455489
11	0.0215023760
12	0.0107548935
13	0.0053779101
14	0.0026890130

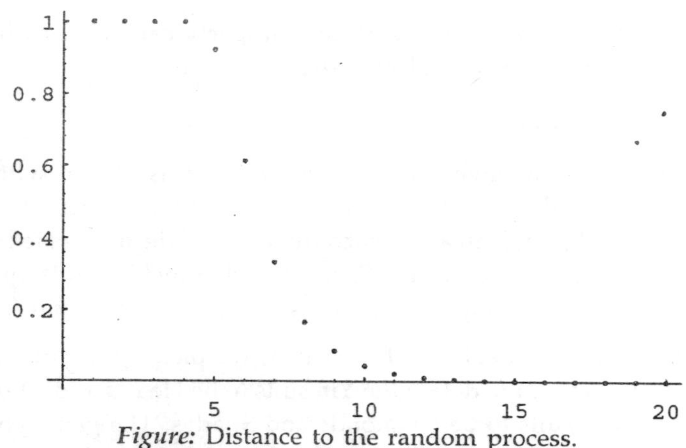

Figure: Distance to the random process.

After 5 shuffles, the distance from the random process is essentially halved each time a shuffle occurs.

Given the distribution functions $f_X(\pi)$ and $u(\pi)$ as above, there is another way to view the variation distance $\|f_X - u\|$. Given any event T (which is a subset of S_n), we can calculate its probability under the process X and under the uniform process. For example, we can imagine that T represents the set of all permutations in which the first player in a 7-player poker game is dealt a straight flush (five consecutive cards in the same suit). It is interesting to consider how much the probability of this event after a certain number of shuffles differs from the probability of this event if all permutations are equally likely. This difference can be thought of as describing how close the process X is to the random process with respect to the event T.

Now consider the event T such that the absolute value of the difference between these two probabilities is as large as possible. It can be shown that this absolute value is the variation distance between the process X and the uniform process. (The reader is asked to prove this fact in Exercise 4.)

We have just seen that, for a deck of 52 cards, the variation distance between the 7-riffle shuffle process and the random process is about .334. It is of interest to find an event T such that the difference between the probabilities that the two processes produce

T is close to .334. An event with this property can be described in terms of the game called New-Age Solitaire.

New-Age Solitaire

This game was invented by Peter Doyle. It is played with a standard 52-card deck. We deal the cards face up, one at a time, onto a discard pile. If an ace is encountered, say the ace of Hearts, we use it to start a Heart pile. Each suit pile must be built up in order, from ace to King, using only subsequently dealt cards.

Once we have dealt all of the cards, we pick up the discard pile and continue. We define the Yin suits to be Hearts and Clubs, and the Yang suits to be Diamonds and Spades. The game ends when either both Yin suit piles have been completed, or both Yang suit piles have been completed. It is clear that if the ordering of the deck is produced by the random process, then the probability that the Yin suit piles are completed first is exactly 1/2.

Now suppose that we buy a new deck of cards, break the seal on the package, and riffle shuffle the deck 7 times. If one tries this, one finds that the Yin suits win about 75 per cent of the time. This is 25 per cent more than we would get if the deck were in truly random order. This deviation is reasonably close to the theoretical maximum of 33.4 per cent obtained above.

Why do the Yin suits win so often? In a brand new deck of cards, the suits are in the following order, from top to bottom: ace through King of Hearts, ace through King of Clubs, King through ace of Diamonds, and King through ace of Spades. Note that if the cards were not shuffled at all, then the Yin suit piles would be completed on the first pass, before any Yang suit cards are even seen. If we were to continue playing the game until the Yang suit piles are completed, it would take 13 passes through the deck to do this.

Thus, one can see that in a new deck, the Yin suits are in the most advantageous order and the Yang suits are in the least advantageous order. Under 7 riffle shuffles, the relative advantage of the Yin suits over the Yang suits is preserved to a certain extent.

Exercises

1. Given any ordering σ of {1, 2, . . ., n}, we can define σ^{-1}, the inverse ordering of σ, to be the ordering in which the ith element is the position occupied by i in ¾. For example, if σ = (1, 3, 5, 2, 4, 7, 6), then σ^{-1} = (1, 4, 2, 5, 3, 7, 6). (If one thinks of these orderings as permutations, then σ^{-1} is the inverse of σ).

   A fall occurs between two positions in an ordering if the left position is occupied by a larger number than the right position. It will be convenient to say that every ordering has a fall after the last position. In the above example, σ^{-1} has four falls. They occur after the second, fourth, sixth, and seventh positions. Prove that the number of rising sequences in an ordering σ equals the number of falls in σ^{-1}.

2. Show that if we start with the identity ordering of {1, 2, . . ., n}, then the probability that an a-shuffle leads to an ordering with exactly r rising sequences equals

$$\frac{\binom{n+a-r}{n}}{a^n}A(n,r),$$

for $1 \le r \le a$.

3. Let D be a deck of n cards. We have seen that there are a^n a-shuffles of D. A coding of the set of a-unshuffles was given in the proof of Theorem (i). We will now give a coding of the a-shuffles which corresponds to the coding of the a-unshuffles. Let S be the set of all n-tuples of integers, each between 0 and $a - 1$. Let $M = (m_1, m_2, \ldots, m_n)$ be any element of S. Let n_i be the number of i's in M, for $0 \le i \le a - 1$. Suppose that we start with the deck in increasing order (i.e., the cards are numbered from 1 to n). We label the first n_0 cards with a 0, the next n_1 cards with a 1, etc. Then the a-shuffle corresponding to M is the shuffle which results in the ordering in which the cards labelled i are placed in the positions in M containing the label i. The cards with the same label are placed in these positions in increasing order of their numbers.

For example, if $n = 6$ and $a = 3$, let $M = (1, 0, 2, 2, 0, 2)$. Then $n_0 = 2$, $n_1 = 1$, and $n_2 = 3$. So we label cards 1 and 2 with a 0, card 3 with a 1, and cards 4, 5, and 6 with a 2. Then cards 1 and 2 are placed in positions 2 and 5, card 3 is placed in position 1, and cards 4, 5, and 6 are placed in positions 3, 4, and 6, resulting in the ordering (3, 1, 4, 5, 2, 6).

(a) Using this coding, show that the probability that in an a-shuffle, the first card (i.e., card number 1) moves to the ith position, is given by the following expression:

$$\frac{(a-1)^{i-1} a^{n-i} + (a-2)^{i-1} (a-1)^{n-i} + \cdots + 1^{i-1} 2^{n-i}}{a^n}.$$

(b) Give an accurate estimate for the probability that in three riffle shuffles of a 52-card deck, the first card ends up in one of the first 26 positions. Using a computer, accurately estimate the probability of the same event after seven riffle shuffles.

4. Let X denote a particular process that produces elements of S_n, and let U denote the uniform process. Let the distribution functions of these processes be denoted by f_X and u, respectively. Show that the variation distance $\|f_X - u\|$ is equal to

$$\max_{T \subset S_n} \sum_{\pi \in T} (f_X(\pi) - u(\pi))$$

Hint: Write the permutations in S_n in decreasing order of the difference $f_X(\pi) - u(\pi)$.

5. Consider the process described in the text in which an n-card deck is repeatedly labelled and 2-unshuffled. The process continues until the labels are all different. Show that the process never terminates until at least $\lceil \log_2(n) \rceil$ unshuffles have been done.

Bibliography

Aitchinson, J. and Brown J.A.C.: *The Lognormal Distribution*, Cambridge University Press, 1957.

Aleong, J. and Haugh L. G.: *Total Quality Management: A New Course for the Statistics Curriculum*, American Statistical Association, Massachusetts, 1992.

Anderson, C. W. and Loynes R. M.: *The Teaching of Practical Statistics*, John Wiley and Sons, New York, 1987.

Andrews, D. M. and Herzberg A.: *Data: A Collection of Problems from Many Fields for the Student and Research Workers*, Springer-Verlag, New York, 1985.

Bancroft, T. A. and Huntsberger D. V.: *Some Papers Concerning the Teaching of Statistics*, Iowa State University Press, Iowa, 1961.

Barnard, G.A.: *Pivotal Inference and the Bayesian Controversy*, Bull. Int. Statist, 1977.

Barndorff-Nielsen O. and Sato K.: *Aarhus Lecture Notes on Stochastic Processes, Lecture Notes Series 16*, Aarhus University, Denmark, 2004,

————————: *Stochastic Processes*, Aarhus University, 1969.

Barndor-Nielsen, O.E.: *Hyperbolic Distributions and Ramications: Con- tribution to Theoryand Application*, Researchreport University of Aarhus, 1980.

Barnett, V.: *Advanced Level Studies: Statistics, Statistical Background*, University of Sheffield, Sheffield, 1985.

―――――――: *Teaching Statistics in Schools Throughout the World*, International Statistical Institute, The Netherlands, 1982.

Baron, J. S. and Worsdale G. J.: *Assignments in Quantitative Methods*, Polytech Publications, Cheshire, 1980.

Barut A. O.: *Foundations of Radiation Theory and Quantum Electrodynamics*, Plenum Press, New York, 1980.

Bayes, T.: *An EssayTowards Solving a Problem in the Doctrine of Chances*, Philosophical Transactions of the Royal Society of London, 1764.

Bennet, G.W. and Cornish E.A.: *A Comparison of the Simultaneous Fidu-cial Distribution Derived from the Multivariate Normal Distribution*, In- tern. Statist. Inst. Bull., 1963.

Berger, J.O.: *Statistical Decision Theory and Bayesian Analysis*, Springer Verlag, 1980.

Bernardo, J. M. Lindly D. V. and Smith A. F. M.: *What is the Question?*, Valencia Univ. Press, Valencia, 1987.

Bhattacharyya, R. K.: *Teaching Probability and Statistics in the First Degree Level Mathematics Major: A Review*, University of Otago, New Zealand, 1990.

Bibby, J.: *History of Teaching Statistics*, John Bibby, London, 1986.

Boen, J. R. and Zahn D. A.: *The Human Side of Statistical Consulting*, Lifetime Learning Publications, California, 1982.

Bowen, J. V.: *Results of Visualization Testing in Introductory Statistics Course*, Association, Massachusetts, 1992.

Breny, H.: *The Teaching of Statistics in Schools*, International Statistical Institute, The Netherlands, 1976.

Burrill, G.: *From Home Runs to Housing Costs: Data Resource for Teaching of Statistics*, Dale Seymour Publications, California, 1994.

―――――――: *Guidelines for the Teaching of Statistics*, American Statistical Association, Alexandria, 1991.

Croasdale, R., *Student Projects in Statistics: Proceedings of an ASLIP Conference*, Association of Statistical Lecturers in Polytechnics, Cheshire, 1985.

Dale, Andrew I.: *A History of Inverse Probability: From Thomas Bayes to Karl Pearson,* Springer-Verlag, New York, 1991.

Daston, Lorraine: *Classical Probability in the Enlightenment,* Princeton University Press, Princeton, 1988.

David, F. N.: *Games, Gods and Gambling: The Origins and History of Probability and Statistical Ideas from the Earliest Times to the Newtonian Era,* Griffin, London, 1962.

Davies, N.: *Teaching and Using Statistics,* Royal Statistical Society, London, 1993.

Edgeman, R. L.: *Quality Improvement and Statistical Reasoning: A Business School Implementation,* American Statistical Association, Georgia, 1991.

Ferguson, T.S.: *Mathematical Statistics: A Decision Theoretic Approach,* Academic Press, 1967.

Finkelstein, D. M.: *Teaching Statistical Consultation with Medical Collaborators,* American Statistical Association, Massachusetts, 1992.

Fisher, E.: *Computer Laboratory Assignments in a Mathematical Statistics Course,* American Statistical Association, Boston, 1992.

Fisher, R.A.: *Inverse Probability,* Proceedings of the Cambridge Philosoph- ical Society, 1930.

Flashoff, J. D.: *What Sample Size do I Need?,* American Statistical Association, Massachusetts, 1992.

Friedman A. and Pinsky M.: *Stochastic Analysis in Infinite Dimensions,* Academic Press, New York.

Friel, S. N. and Mokros J. R.: *Used Numbers, Statistics – Middles, Means and in between,* Dale Seymour Publications, California, 1990.

Fuller, M.: *Problems of Learning Statistics,* Sheffield City Polytechnic, Sheffield, 1980.

Gal, I. and Garfield J.: *The Assessment Challenge in Statistics,* International Statistical Institute, The Netherlands, 1997.

Goldsmith, C. A.: *Teaching Critical Appraisal Principles,* American Statistical Association, Massachusetts, 1992.

Graham, A. *Teach Yourself Statistics,* Hodder and Soughton, London, 1994.

Grandy, W. T. and Milonni P. W.: *Physics and Probability,* Cambridge Univ. Press, Cambridge, 1993.

Green, D. R.: *Probability Concepts in 11-16 Year Old Pupils,* Loughborough University, Loughborough, 1982.

——————: *Teaching Statistics at its Best,* Teaching Statistics Trust, Sheffield, 1993.

Greenspan, P. and Fisch G. S.: *Visual Inspection of Data: A Behavioral Analysis,* American Statistical Association, Massachusetts, 1992.

Habibullah, S. N.: *Efforts to Promote Project-Based Teaching of Introductory Statistics in the City of Lahore,* Laval University, Canada, 1992.

——————: *Proposal for a Comprehensive Four-Year Program of Study in Statistics at the Undergraduate Level in Pakistan,* Allama Iqbal Open University, Pakistan, 1990.

——————: *Survey Work to Enhance Understanding of Statistics in BA/BSc Students,* University Technology Malaysia, Malaysia, 1990.

Hacking, Ian: *The Emergence of Probability: A Philosophical Study of Early Ideas about Probability, Induction and Statistical Inference,* Cambridge University Press, New York, 1975.

Hald, Anders: *A History of Probability and Statistics and their Applications before 1750,* Wiley, New York, 1990.

Hamzeh, A.: *Teaching Statistics by Objectives,* American Statistical Association, Massachusetts, 1992.

Hand, D. J. and Everitt B.: *The Statistical Consultant in Action*, Cambridge University Press, Cambridge, 1987.

Hand, D. J. and Ostrowski E.: *A Handbook of Small Data Sets*, Chapman and Hall, London, 1994.

Hawkins, A.: *Training Teachers to Teach Statistics*, International Statistical Institute, The Netherlands, 1990.

Heidelberger, Michael and Rosemarie Rheinwald: *Probability since 1800: Interdisciplinary Studies of Scientific Development*, B. Kleine Verlag, Bielefeld, 1983.

Heyde, C. C. and Seneta, E.: *Bienayme: Statistical Theory Anticipated*, Springer-Verlag, New York, 1977.

Hildebrand, D. K.: *Making Statistics More Effective in School of Business: Improving Statistics Textbooks*, American Statistical Association, Atlanta, 1991.

Hilts, Victor L.: *Statist and Statistician*, Arno Press, New York, 1981.

Holmes, P.: *Teaching Statistics 11-16*, Foulsham Educational, Slough, 1980.

Howard, K. and Sharp J. A.: *The Management of a Student Research Project*, Gower Press, Aldershot, 1983.

Husein, A. and Slotnick H. B.: *A Note on Teaching Experimental Design and Linear Models*, American Statistical Association, Massachusetts, 1992.

Jaynes, E. T.: *Proceedings of the Second Rochester Conference on Coherence and Quantum Optics*, Plenum, New York, 1966.

——————: *Quantum Beats*, Plenum Press, New York, 1980.

——————: *Where do we Stand on Maximum Entropy?*, M. I. T. Press, Cambridge, 1979.

Jereys, H.: *Scientic Inference*, Cambridge University Press, 1957.

——————: *Theory of Probability*, Oxford University Press, 1939.

Jolliffe, F. and Glickman L.: *Teaching Statistical Concepts*, Longman, London, 1992.

Justice J. H.: *Maximum-Entropy and Bayesian Methods in Applied Statistics,* Cambridge Univ. Press, Cambridge, 1986.

Kader, G.: *Implementing the NCTM Standards in Probability and Statistics in Grades 9-12,* American Statistical Association, 1992.

Kane, V. E.: *The Role Statistics Can Play in Competitive Management,* American Statistical Association, Georgia, 1991.

Konold, C.: *Chance Plus: A Computer Based Curriculum for Probability and Statistics,* University of Massachusetts, Massachusetts, 1990.

Lachenbruch, P. A.: *Bringing Computers to Class,* American Statistical Association, Massachusetts, 1992.

Levine, R. D. and Tribus M.: *The Maximum Entropy Formalism,* M. I. T. Press, Cambridge, 1979.

Lind, R. B.: *Data, Concepts and Communication in a Math Stat Course,* American Statistical Association, Boston, 1992.

Maistrov, L. E.: *Probability Theory: A Historical Sketch,* Academic Press, New York, 1974.

Moore G. T. and Scully M. O.: *Frontiers of Nonequilibrium Statistical Physics,* Plenum Press, New York, 1986.

Moore, D. S. and Hoaglin D.: *Perspectives on Contemporary Statistics,* Mathematical Association of America, Washington, 1992.

O'Hear, A.: *Popper's Contribution to the Philosophy of Probability,* Cambridge University Press, 1995.

Owen, D. B.: *On the History of Statistics and Probability: Proceedings of A Symposium on the American Mathematical Heritage,* Dekker, New York, 1976.

Pearson, E. S.: *Studies in the History of Statistics and Probability: A Series of Papers,* Griffin, London, 1970.

Porter, Theodore M.: *The Rise of Statistical Thinking, 1820-1900,* Princeton University Press, Princeton, 1986.

Rade, L. and Speed T.: *Teaching of Statistics in the Computer Age,* Chartwell-Bratt Ltd., Kent, 1985.

Rustagi, J. S. and Wolfe D. A.: *Teaching of Statistics and Statistical Consulting,* Academic Press, New York, 1982.

Shulte, A. and Smart J. R.: *Teaching Statistics and Probability,* National Council of Teachers of Mathematics, Virginia, 1981.

Stigler, S.M.: *The Measurement of Uncertainty before 1900,* Harvard University Press, 1986.

——————: *The History of Statistics: The Measurement of Uncertainty before 1900,* Belknap Press of Harvard University Press, Cambridge, 1986.

Stochastic, Integration: *Vector and Operator Valued Measures and Applications,* Academic Press, New York,

Tankard, James W.: *The Statistical Pioneers,* Schenkman, Cambridge, 1984.

Todhunter, Isaac: *History of the Mathematical Theory of Probability from the time of Pascal to that of Laplace,* MacMillan, London, 1865.

Von Neumann J. and Morgenstern O.: *Theory of Games and Economic Behaviour,* Princeton University Press, 1947.

Watkins, A. and Landwehr J. M.: *Exploring Data,* Dale Seymour Publications, California, 1986.

Watkins, A. and Landwehr J. M.: *Exploring Surveys and Information from Samples,* Dale Seymour, California, 1987.

Yushkevich A. P.: *Mathematics of the 19th Century: Mathematical Logic, Algebra Number theory, Probability Theory,* Birkhauser, Boston, 1992.

Index

A

B

C

D

E

F

G

Q

R

S

T

U

V

W

Z

□□□